STRATEGIC STUDIES INSTITUTE

The Strategic Studies Institute (SSI) is part of the U.S. Army War College and is the strategic-level study agent for issues related to national security and military strategy with emphasis on geostrategic analysis.

The mission of SSI is to use independent analysis to conduct strategic studies that develop policy recommendations on:

- Strategy, planning, and policy for joint and combined employment of military forces;

- Regional strategic appraisals;

- The nature of land warfare;

- Matters affecting the Army's future;

- The concepts, philosophy, and theory of strategy; and,

- Other issues of importance to the leadership of the Army.

Studies produced by civilian and military analysts concern topics having strategic implications for the Army, the Department of Defense, and the larger national security community.

In addition to its studies, SSI publishes special reports on topics of special or immediate interest. These include edited proceedings of conferences and topically oriented roundtables, expanded trip reports, and quick-reaction responses to senior Army leaders.

The Institute provides a valuable analytical capability within the Army to address strategic and other issues in support of Army participation in national security policy formulation.

ARMY SUPPORT OF MILITARY CYBERSPACE OPERATIONS: JOINT CONTEXTS AND GLOBAL ESCALATION IMPLICATIONS

Jeffrey L. Caton

January 2015

*****

Comments pertaining to this report are invited and should be forwarded to: Director, Strategic Studies Institute and U.S. Army War College Press, U.S. Army War College, 47 Ashburn Drive, Carlisle, PA 17013-5010.

*****

This manuscript was funded by the U.S. Army War College External Research Associates Program. Information on this program is available on our website, *www.StrategicStudies Institute.army.mil*, at the Opportunities tab.

*****

*****

The Strategic Studies Institute and U.S. Army War College Press publishes a monthly email newsletter to update the national security community on the research of our analysts, recent and forthcoming publications, and upcoming conferences sponsored by the Institute. Each newsletter also provides a strategic commentary by one of our research analysts. If you are interested in receiving this newsletter, please subscribe on the SSI website at *www.StrategicStudiesInstitute.army.mil/newsletter*.

ISBN 1-58487-660-3

FOREWORD

Military cyberspace operations have been ongoing since before the advent of the Internet, and their influence on traditional military operations continues to increase. What are the significant changes in mission and structure of Department of Defense offensive and defensive cyberspace activities over the past decade? How do joint and Army cyberspace military operations fit into the complex and dynamic sphere of daily network defense as well as international deterrence and escalation?

To facilitate the operationalization of this new domain, education of the tenets of cyberspace must occur at the tactical, operational, and strategic levels of leadership. The persistent increase of cyberspace activities in global events continues to make international dynamics more complex. The scope of context for such matters needs to consider not just other military efforts or even other instruments of national power, but how they are presented in an escalation framework and where they may be going.

This monograph posits that expanding deterrence forces to include conventional strike and cyber offense can add capability and credibility, as well as flexibility, to course-of-action development available for national command authorities. It also argues that cyberspace operations, such as automated cyber defense, can support and enhance deterrence operations and limited conflict as well as help control escalation and reduce risk.

DOUGLAS C. LOVELACE, JR.
Director
Strategic Studies Institute and
    U.S. Army War College Press

ABOUT THE AUTHOR

JEFFREY L. CATON is President of Kepler Strategies LLC, Carlisle, PA, a veteran-owned small business specializing in national security, cyberspace theory, and aerospace technology. He is also an Intermittent Professor of Program Management with the Defense Acquisition University. From 2007-12, Mr. Caton served on the U.S. Army War College faculty, including Associate Professor of Cyberspace Operations and Defense Transformation Chair. Over the past 5 years, he has presented lectures on cyberspace and space issues related to international security in the United States, Sweden, the United Kingdom, Estonia, and Kazakhstan, supporting programs such as the Partnership for Peace Consortium and the North Atlantic Treaty Organization Cooperative Cyber Defence Center of Excellence. His current work includes research on cyberspace and space issues as part of the External Research Associates Program of the Strategic Studies Institute as well as serving as a facilitator for Combined/Joint Land Force Component Commander courses at the Center for Strategic Leadership and Development. He served 28 years in the U.S. Air Force working in engineering, space operations, joint operations, and foreign military sales, including command at the squadron and group level. Mr. Caton holds a bachelor's degree in chemical engineering from the University of Virginia, a master's degree in aeronautical engineering from the Air Force Institute of Technology, and a master's degree in strategic studies from the Air War College.

SUMMARY

Military cyberspace operations have been ongoing since before the advent of the Internet. Such operations have evolved significantly over the past 2 decades and are now emerging into the realm of military operations in the traditional domains of land, sea, and air. The goal of this monograph is to provide senior policymakers, decisionmakers, military leaders, and their respective staffs with a better understanding of Army cyberspace operations within the context of overall U.S. military cyberspace operations. It first looks at the evolution of Department of Defense (DoD) cyberspace operations over the past decade. Next, it examines the evolution of the Army implementation of cyberspace operations. Finally, it explores the role of cyberspace operations in the escalation of international conflict.

The scope of discussion is at the survey level of detail to provide an overall appreciation for the complex and dynamic nature of evolving cyberspace operations. It is limited to unclassified and open source information; any classified discussion must occur at an appropriate venue. Although the details contained herein are largely focused on military applications, the reader must realize that whole-of-government efforts are essential for the successful implementation of national security efforts in cyberspace.

This monograph has three main sections:

- Evolution of Military Cyberspace Operations. This section examines the founding of U.S. Cyber Command from its roots in various military units focused on defensive and offensive cyberspace operations. It reviews the initial operation of the command under the leadership of General Keith Alexander as well as its current operations led by Admiral Michael Rogers. Also, it assesses the command's mission to direct operations, defend networks, and, on order, conduct full spectrum operations, with respect to its appropriateness and adeptness for the command and control of military cyberspace forces.

- Evolution of Army Cyberspace Operations. Having examined the evolution of joint cyberspace operations, this section focuses on parallel evolutionary efforts in Army cyberspace operations toward the establishment of Army Cyber Command. It examines initial operations of the command under the leadership of Lieutenant General Rhett Hernandez as well as its current operations led by Lieutenant General Edward Cardon. This includes a brief review of recent efforts to establish Fort Gordon, Georgia as the center of gravity for Army cyberspace activities.

- Cyberspace Operations in a Global Context. This section examines the sufficiency of the current cyberspace force structure to address an international environment of multiple actors interacting with varying degrees of tension. In such a global situation, cyberspace operations seeking to produce certain effects must also be examined for their potential to cause escalation of activities; possibly even up to the point of existential threat. The section presents a modified Kahn escalation ladder as a useful metaphor to explore how cyberspace activities may integrate with traditional military operations across the spectrum of international conflict as well as how such defenses influence national responses related to deterrence and escalation.

This monograph examines the past and present joint and Army cyberspace military operations, as well as how these operations may fit into the complex and dynamic sphere of international deterrence and escalation. To facilitate the best evolutionary path for future activities, it provides recommendations in the areas of current priorities, authorities, strategic engagement, multi-role modeling, and other paradigms and factors to consider in future examinations of the topic.

AUTHOR'S NOTE

When this monograph was initially completed in August 2012, the capstone doctrine document for U.S. military cyberspace operations—*Joint Publication (JP) 3-12, Joint Cyberspace Operations*—was a classified document. On October 21, 2014, the Joint Chiefs of Staff released *JP 3-12(R), Cyberspace Operations*, an unclassified version of the earlier doctrine document that is posted on the unclassified public access government website "Joint Electronic Library" (available from *www.dtic.mil/doctrine/*). Please note that the cover of the unclassified version retains the original classified release date of February 5, 2013, but its contents do not include an explanatory note as to when, how, and why this declassification was made.

In general terms, the information in this monograph is consistent with the details contained in *JP 3-12(R)*, and thus this monograph has not been modified to assess and incorporate this recent release. However, a diagram from *JP 3-12 (R)* that depicts typical joint cyberspace command and control organizational relationships is included as Figure A-1 in the Appendix to complement the information contained in Figures 1, 2, and 3 of this monograph.

ARMY SUPPORT OF MILITARY CYBERSPACE OPERATIONS: JOINT CONTEXTS AND GLOBAL ESCALATION IMPLICATIONS

Military cyberspace operations have been ongoing since before the advent of the Internet. Such operations have evolved significantly over the past 2 decades and are now emerging into the realm of military operations in the traditional domains of land, sea, and air. The goal of this monograph is to provide senior policymakers, decisionmakers, military leaders, and their respective staffs with a better understanding of Army cyberspace operations within the context of overall U.S. military cyberspace operations. To accomplish this, it first looks at the evolution of Department of Defense (DoD) cyberspace operations over the past decade. Next, it examines the evolution of the Army implementation of cyberspace operations. Finally, it explores the role of cyberspace operations in the escalation of international conflict. The scope of discussion is at the survey level of detail to provide an overall appreciation for the complex and dynamic nature of evolving cyberspace operations. It is limited to unclassified and open source information; any classified discussion must occur at an appropriate venue. Although the details contained herein are largely focused on military applications, the reader must realize that whole-of-government efforts are essential for the successful implementation of national security efforts in cyberspace.

EVOLUTION OF MILITARY CYBERSPACE OPERATIONS

This section examines the founding of the U.S. Cyber Command from its roots in various military units focused on defensive and offensive cyberspace operations. It reviews the initial operation of the command under the leadership of General Keith Alexander as well as its current operations led by Admiral Michael Rogers. Also, it assesses the command's mission to direct operations, defend networks, and, on order, conduct full spectrum operations with respect to its appropriateness and adeptness for the command and control of military cyberspace forces.

The Founding of U.S. Cyber Command.

The formal establishment of military units dedicated to cyberspace missions is mostly a phenomenon of the 21st century. This section will look at how the defensive and offensive aspects of cyberspace operations evolved until they were merged under U.S. Cyber Command.

Defensive Cyberspace: Joint Task Force-Global Network Operations.

In the last years of the 20th century, DoD began to form the forerunners of a dedicated cyberspace force. In December 1998, Secretary of Defense William Cohen approved formation of the Joint Task Force-Computer Network Defense (JTF-CND) to "serve as the focal point with the Department of Defense to organize a united effort to defend its computer networks and systems" based on needs demonstrated by "defense exercises and real world events in 1997 and in early 1998."[1] These events included the DoD Eligible Receiver 1997 exercise as well as the hacking incidents known as Solar Sunrise and Moonlight Maze.[2] JTF-CND was collocated with the Global Operations and Security Center of the Defense Information Systems Agency (DISA) in Washington, DC, and was given the initial mission to be responsible for operations on DoD computer systems and networks as well as coordinating these efforts with the interagency and commercial communities.[3]

The initial cadre was small at 10 personnel assigned and only 24 assigned when full operational capability was achieved in June 1999. At first, the JTF-CND was not assigned to a unified command, so its commander reported through the Chairman of the Joint Chiefs of Staff to the Secretary of Defense.[4] The first commander, Major General John Campbell, recognized there was no connection with services and regional warfighting commanders, and the interim command arrangement evolved quickly.[5] Within a year, JTF-CND was placed under the U.S. Space Command with responsibilities that included DoD-wide defense actions to stop computer network attack (CNA) and computer network exploitation (CNE) efforts and to mitigate the effects of any successful attacks.[6]

In April 2001, the offensive cyberspace role of computer network attack was assigned to U.S. Space Command, and the JTF-CND was renamed to Joint Task Force-Computer Network Operations (JTF-CNO).[7] The new commander, Major General James Bryan, was also dual-hatted as Vice Director, DISA. He described the new organization and reporting structure to Congress in May 2001:

Sir, Joint Task Force-CNO is, in fact, that one-stop operational command for the Department of Defense for both offense and defense. It is important to remember that we may be a one-stop shop for operational coordination; but without the cooperation of the services and the agencies to include law enforcement as part of one team, the JTF could not do its job as well as we do. But it certainly answers the question as to who is in charge, and this operational accountability now flows from the President to the Secretary of Defense to General Eberhardt, who is CINCSPACE, to me.[8]

On January 10, 2003, President George W. Bush signed Change-2 to the 2002 Unified Command Plan, which included the merging of U.S. Space Command and the existing U.S. Strategic Command into the "new" U.S. Strategic Command (USSTRATCOM) under which JTF-CNO was realigned.[9] By April 2004, the first Concept of Operations for network operations (NetOps) for the DoD global information grid (GIG) was approved. The roles of defensive and offensive cyberspace activities were refined during this period such that in July 2004, Secretary of Defense Donald Rumsfeld changed JTF-CNO to Joint Task Force-Global Network Operations (JTF-GNO).[10] The first JTF-GNO commander was the director of DISA, Lieutenant General Harry Raduege, Jr., who later noted:

For the first time in network operations and cybersecurity history, command lines were established from the secretary of defense to the STRATCOM commander, to the JTF-GNO commander, to each of the appointed component commanders within the military services and representatives within the combatant commands and defense agencies. This framework provides an important governance model for optimally operating and defending Defense Department networks through an established command structure.[11]

After the inaugural year of operations, USSTRAT-COM commander, General James Cartwright, approved a revised Concept of Operations (CONOPS) to capture lessons learned for JTF-GNO on August 15, 2005. The CONOPS noted that the NetOps primary mission to operate and defend the DoD's critical information backbone—the GIG—is explicitly an ongoing one: "Unlike many missions that are deemed successful at a defined completion date, the NetOps mission is perpetual, requiring continual support to be successful."[12] To accomplish this, the CONOPS envisioned six critical capabilities to be employed across the spectrum of DoD operations at the strategic, operational, and tactical levels: visibility; monitoring and analysis; planning; coordinating and responding; management and administration; and control.[13]

Some of the practical aspects of the revised CONOPS were its delineation of NetOps within the context of joint and Service organizations. It also distinguished between NetOp events (activities that may impact operational readiness of the GIG) at the theater level and global level. NetOps Events with effects limited to a specific theater's operations—Theater NetOp Events—would be under the control of the appropriate geographic commander in the supported role, receiving necessary support from USSTRATCOM and JTF-GNO. For NetOps Events with the potential to impact the GIG across multiple theaters—Global NetOps Events—the commander, USSTRATCOM, would be the supported commander and would issue orders through to JTF-GNO to combatant commands, services, and agencies via established support and command centers.[14]

The command and control structure for addressing NetOps Events utilized NetOps Control Centers at the theater level (TNCC), global level (GNCC), and joint level (JNCC). The CONOPS called for TNCCs at U.S. Central Command, U.S. European Command, U.S. Northern Command, U.S. Pacific Command, and U.S. Southern Command:

> to lead, prioritize, and direct Theater GIG assets and resources to ensure they are optimized to support the GCC's [geographic combatant command's] assigned missions and operations, and to advise the COCOM [combatant command] of the ability of the GIG to support current and future operations.[15]

As part of their Global NetOps Event responsibilities, a GNCC would provide support to functional combatant commands (FCCs), such as U.S. Transportation Command "to advise the FCC and ensure the portion of the GIG resources supporting that Commander's assigned missions and operations are optimized."[16]

The CONOPS also had service and interagency provisions as well as JNCCs to support a joint task force (JTF) commander by managing "the tactical communications of the joint force, serving as the NOSC [Network Operations and Security Center] for the deployed portion of the GIG supporting a JTF."[17] To orchestrate all of these functions, the JTF-GNO commander was assigned several critical responsibilities to ensure proper operation and defense of the GIG, which in turn supported the missions of combatant commands, services, and agencies as well as those of the President and Secretary of Defense.[18]

Finally, the CONOPS set the expectation and measure of merit for its support to the warfighter simply as "the effectiveness of NetOps will be measured in terms of availability and reliability of net-centric services, across all domains, in adherence to agreed-upon service levels and policies."[19] The tenets of the 2005 CONOPS continued to mature through daily operations for several years pursuing a challenge that was conveyed in the December 2008 DoD NetOps Strategic Vision, which strived for the GIG to operate "as a single, unified, agile, and adaptive enterprise capable of providing responsive and resilient support to multiple simultaneous mission areas under uncertain and changing conditions."[20] To address this challenge, the DoD Chief Information Officer set three goals: share GIG situational awareness; unify GIG command and control; and institutionalize NetOps.[21] Also, the broad responsibilities regarding NetOps for combatant commands expressed in the USSTRATCOM CONOPS were formally institutionalized as an integral part of the GIG by DoD that month as well.[22]

Offensive Cyberspace: Joint Functional Component Command-Network Warfare.

In 2003, around the same time that JTF-CNO was adjusting its organization to the reporting chain in USSTRATCOM, the DoD offensive cyberspace mission of network attack was transferred to a Network Attack Support Staff also under the operational control of USSTRATCOM but collocated with the National Security Agency (NSA).[23] By January 2005, this staff evolved to become the Joint Functional Component Command—Network Warfare (JFCC-NW).[24] The Director of the NSA was designated as the commander of JFCC-NW and thus the offensive cyberspace mission was separated from the defensive cyberspace mission carried out by the Director of DISA in the role of commander, JTF-GNO.[25] The 2005 USSTRATCOM

NetOps CONOPS defined the primary responsibilities of JFCC-NW as "planning, integrating and coordinating computer network warfare capabilities and integrating with all necessary computer network defense and exploitation capabilities."[26]

Further details of the capabilities and implementation of offensive cyberspace operations remain classified. For public dissemination, Lieutenant General Keith Alexander (Director, NSA and commander, JFCC-NW) summed up the state of cyberspace operations in a 2007 article as:

> We [USSTRATCOM] have redefined our cyberspace mission area in terms of offensive—network warfare (NW) and defensive—network operations (NetOps)—and established JFCC-NW and JTF-GNO to address each of those mission sets, respectively.
>
> USSTRATCOM has also begun to develop tactics, techniques, and procedures and other concepts designed to integrate cyberspace capabilities into cross-mission strike plans. We are developing concepts to address warfighting in cyberspace in order to assure freedom of action in cyberspace for the United States and our allies while denying adversaries and providing cyberspace-enabled effects to support operations in other domains. These concepts, and the cyberspace effects that they focus on, are clearly based on the military concepts of strike, fires (supporting and suppressing), and defense.[27]

This arrangement of two three-star general commanders reporting separately to USSTRATCOM was streamlined in late-2008 when operational command of JTF-GNO was placed under JFCC-NW.[28] This change was intended to "close the seams between information assurance, network operations and defense, intelligence collection and offensive operations."[29]

In the fall of 2010, the world learned of a previously classified cyberspace operation through an article in *Foreign Affairs* by Deputy Secretary of Defense William J. Lynn III. Calling the 2008 incident "the most significant breach of U.S. military computers ever," Lynn went on to note that "the Pentagon's operation to counter the attack, known as Operation BUCKSHOT YANKEE, marked a turning point in U.S. cyber-defense strategy."[30] Part of this strategy included the formation of a new sub-unified command under USSTRATCOM—U.S. Cyber Command (US-CYBERCOM).[31] The creation of USCYBERCOM was directed in a June 23, 2009, memorandum by Secretary of Defense Robert Gates. The new command would incorporate the existing elements of DoD cyberspace such as service component and agency connections. In doing this, Gates also directed the disestablishment of JTF-GNO and JFCC-NW as their functions were subsumed into USCYBERCOM.[32]

The first commander of USCYBERCOM, General Keith Alexander, in testimony to Congress in September 2010, recapped the events from Operation BUCK-SHOT YANKEE up through initial operational capability of the new command as well as how its structure would greatly enhance future cyberspace operations.

> At that time [2008], we had the defense and the operations in one command, under the Joint Task Force-Global Network Operations. And that task force got one level of intelligence and could see one part of the network.
>
> Operating on the other side was the Joint Functional Component Command-Net Warfare, trained at a dif-

ferent level with different intel insights at a different classification level, same network, two organizations. And if you are operating at the National Training Center, you wouldn't have the defensive team out there defending and then take them off the field and run out with an offensive team. It is the same team.

And so the good thing that we have done here is we have brought those two together, merged those, and I think that is key to the success here. We need that to operate as one team. The offense and defense cannot be different here, because these operations will occur in real time. And I think we have to be prepared to do that.[33]

Initial USCYBERCOM Operations.

Secretary of Defense Gates set very aggressive dates for USCYBERCOM establishment: initial operating capability by October 2009 and full operational capability by October 2010.[34] Although the first operational milestone was not achieved until May 21, 2010, USCYBERCOM was declared fully operational, which included the formal disestablishment of JTF-GNO and JFCC-NW.[35] The USCYBERCOM mission was three-fold: enable DoD network operations; conduct military cyberspace operations; and ensure freedom of action in cyberspace.[36]

Figure 1 depicts the interim structures of the developing USCYBERCOM within the larger context of DoD cyberspace. Working in parallel to the joint efforts, each military service was also tasked to develop and establish cyberspace commands to support US-CYBERCOM. By October 2010, the following component support commands were in place: Army Cyber Command; Fleet Cyber Command, 10th Fleet; Marine Forces Cyber; and 24th Air Force.

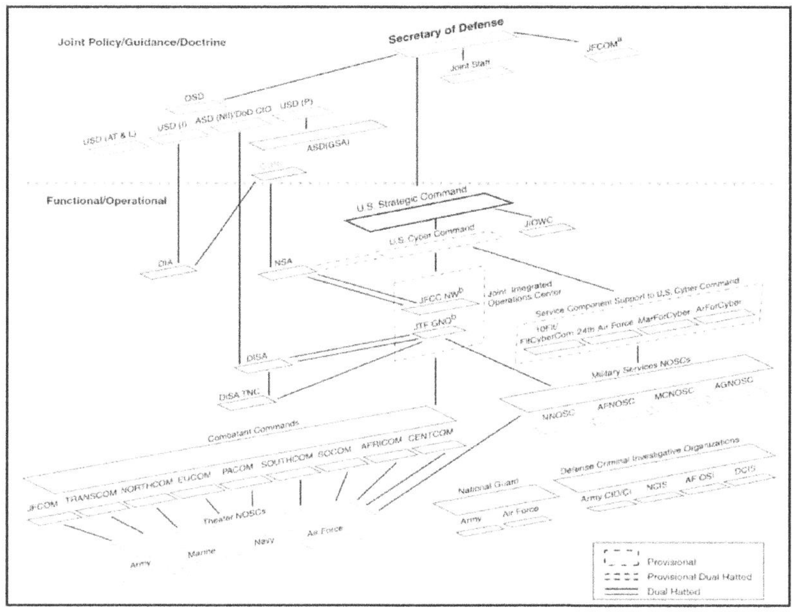

Figure 1. USCYBERCOM Formation and DoD Cyber Organization (March 2010).[37]

Consistent with the vision put forth in the *Foreign Affairs* article by Deputy Secretary Lynn, General Alexander confirmed the initial direction of the first USCYBERCOM was set in three main lines of operation: defense of the Global Information Grid; execution of full-spectrum cyber operations on command; and defense of U.S. freedom of action in cyberspace. He also reiterated five principles for the initial strategy of DoD cyberspace:

- Remember that cyberspace is a defensible domain.
- Make our defense active.
- Extend protection to our critical infrastructure.
- Foster collective defenses.
- Leverage U.S. technological advantages.[38]

What was the vision for the practical application of these principles in military terms? General Alexander emphasized that the need for the command to focus on operating jointly in support of the combatant commanders.[39] This cyberspace support to the deployed warfighter was facilitated using Cyber Support Elements (CSEs) for combatant commanders and Expeditionary CSEs (ExCSEs) for joint task force commanders. These teams are scalable in size and composition to best meet mission requirements as well as establish working relationships with the directorates of intelligence (J2), operations (J3), and planning (J5). Regarding ExCSE activities that support ongoing operations, General Alexander testified to Congress in 2010 that:

> Currently, USCYBERCOM has two ExCSEs teams deployed — one in Iraq and one in Afghanistan. The teams consist of five personnel: a team chief (lead planner), a cyber attack planner, a cyber defense planner, and two analysts (cyber and intelligence). USCYBERCOM embeds these teams within the supported Joint Task Force headquarters (typically J3 Directorate — Operations) to enable the delivery of cyber effects in support of the commander's priorities.[40]

The USCYBERCOM commander would also lead the National Security Agency (NSA) and Central Security Service, thus adding in the traditional communities of national security cryptology, signals intelligence, and information assurance into the cyberspace operations mix. Although this puts a great amount of responsibility under the purview of a single leader, General Alexander argued that it made sense for resource stewardship and unity of effort.[41] From a force structure view, this included the incorporation of existing task-specific support teams, such as:

Green Teams to respond to cyber incidents; Blue Teams that provide in-depth review and resolution of cyber events; and Red Teams that emulate adversary procedures against DoD hosts to train defenders and identify vulnerabilities for mitigation.[42]

Current Joint Cyberspace Operations.

In January 2012, President Barack Obama and Secretary of Defense Leon Panetta gave DoD new strategic guidance for sustaining U.S. global leadership in the 21st century. This guidance centered on 10 primary mission areas where *"the Joint Force will need to recalibrate its capabilities and make selective additional investments to succeed,"* which includes efforts to ensure protection and resiliency for cyberspace operations.[43] Under General Alexander's leadership, US-CYBERCOM pursued five broad command priorities to address the mandate: (1) Trained and Ready Cyber Forces; (2) Operational Concept; (3) Global Situational Awareness; (4) Defensible Architecture; and (5) Policies and Procedures to Enable Action.[44]

Admiral Michael S. Rogers assumed command of USCYBERCOM on April 3, 2014, and since then, he has kept the command focused on the same five priorities.[45] In a June 2014 speech, he highlighted how the Joint Information Environment (JIE) will provide a truly defensible network for warfighters once it is fully mature and noted that the JIE structure is currently being implemented in Europe.[46] He also provided details on the planned structures for trained and ready cyber forces. Consistent with the cyber force envisioned in the 2014 *Quadrennial Defense Review*,[47] Admiral Rogers called for a team structure of approximately 6,000 cyber professionals divided into 133 teams across three mission areas: Cyber National Mission Force responsible for depending national critical infrastructure; Cyber Combat Mission Force responsible for cyber support to combatant commanders; and Cyber Protection Forces responsible for operating and defending the DoD information network (DoDIN).[48] Table 1 depicts how these teams might be aggregated to form notional companies, battalions, and squadrons.

Current Cyberspace Mission Forces	
2014 Quadrennial Defense Review 133 Total Teams 6,000 Pax	13 National Mission Teams with 8 National Support Teams
	27 Combat Mission Teams with 17 Combat Support Teams
	18 National Cyber Protection Teams (CPTs)
	24 Service CPTs
	26 Combatant Command and DoD information Network CPTs
National Basic Types of Cyberspace Units (USCYBERCOM, October 2013)	
Cyber National Mission Battalion/Squadron Mission: See, Block, Maneuver in Red and Grey space to deny adversary objectives and, if authorized, strike to destroy the capability.	1 x C2 Element • Provide C2 and management
	5 x Cyber National Mission Teams (CNMT) (64 Pax each) • Base unit for cyber operations • Conduct OCO/DCO/DGO • Sustained and surge operations • Trained, certified, and fights as a team
	5 x Direct Support Teams (DST) (39 Pax each) • Provides direct support to CNMTs • Conduct intel and malware analysis • Perform immediate tool development / modification and access maintenance • Conduct target discovery / analysis • Provide language analysis • Planning and synchronization • NSA initial weight to DTN DSTs, then shifting to CCMD support as capacity grows.

Table 1. Cyberspace Force Presentation.[49]

Cyber Combat Mission **Battalion/Squadron** Mission: Target development <u>in support of CCMD operations plans</u> and, when authorized, the delivery of cyber effects against CCMD targets, followed by assessment of effects. OPCON to CCMDs under current "Transitional" C2 model.	**1 x C2 Element** • Provide C2 and cyber management for CCMD (OPCON)

	1-6 x Cyber Combat Mission Teams (CCMT) (64 Pax each)
	
	    • Base unit for offensive cyber operations
	    • Large Scale ops CCMF has all CCMT specialties, others less
	    • Trained, certified, and fights as a team

	1-2 x Direct Support Teams (DST) (39 Pax each)
	
	    • One DST per 3-5 CCMT
	    • More target region specific skills
	    • Perform immediate tool development / modification and access maintenance
	    • Conduct target discovery and analysis
	    • Provide language analysis
	    • Planning and synchronization

Cyber Protection
 	
Company / Troop
 	
Mission: Defense of the GIG and employing teams to assist outside the GIG when required and authorized.
	2-6 x Cyber Protection Platoons
	
	    • Each Platoon has its own organic C2 element
	    • Each Platoon has 5 squads (see below)
	    • Conduct CND; tips to CNA; Penetration testing
	    • Trained, certified, and operates as a team

	5 x Protection Squads / Platoons
	
	    • Task organized, trained and certified
	    • Assesses Cyber Security Posture
	    • Bolsters Cyber Defenses
	    • Conducts Counter-Cyber Ops
	    • Performs Cyber Threat Emulation (CTE)
	    • Conducts intel and malware analysis

Table 1. Cyberspace Force Presentation. (cont.)

As Cyber National Mission Force teams are being established, their techniques and procedures are also being developed through daily operations and exercises. Many of these exercises require coordination across multiple lines of authority, such as the Cyber Guard 14-1 exercise conducted over 2 weeks in July 2014 "designed to test operational and interagency coordination as well as tactical-level operations to protect, prevent, mitigate and recover from a domestic cyberspace incident." [50]

Cyber Combat Mission Force teams are also refining their methods for providing support to combatant commanders. As depicted in Figure 2, USCYBERCOM CSEs help to coordinate cyber support through joint component commanders, joint task force commanders, and the combatant commander's Joint Cyber Center. Specific operational requests may be in the form of the Cyber Effects Request Format (CERF) process, which "initiates cyber effects planning across all lines of operation."[51] Warfighters may also use a Joint Cyber Strike Request that "sets the timing and tempo to integrate cyber effects/fires with the supported Joint Force Commander's operation."[52] For planning and execution of these requests, "CDRUSCYBERCOM [Commander, USCYBERCOM] deconflicts fires delivered in and through cyberspace."[53]

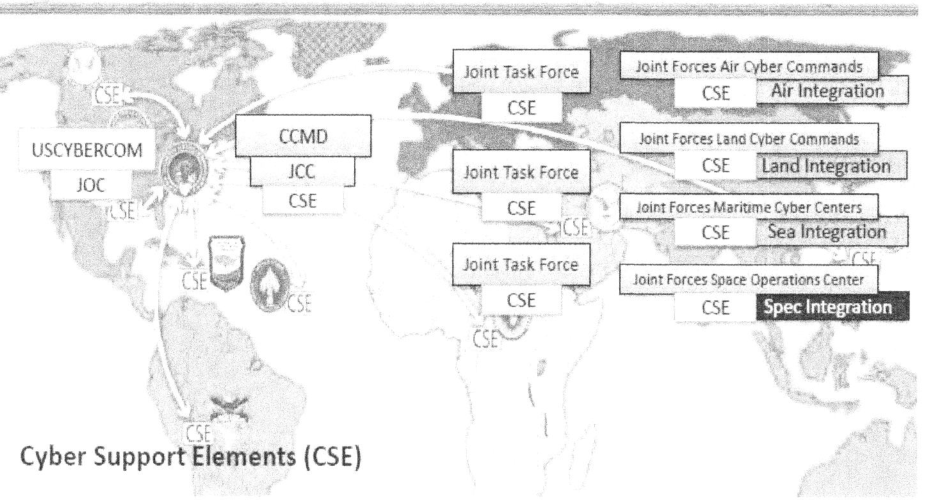

Figure 2. USCYBERCOM Support to Combatant Commands.[54]

From a doctrinal viewpoint, all of the cyberspace operations for warfighters should fall into three mission areas: DoDIN Operations, Defensive Cyberspace Operations (DCO), and Offensive Cyberspace Operations (OCO). DCO is bifurcated into DCO-Internal Defensive Measures (IDM) and DCO-Response Actions (RA).[55] Figure 3 depicts the notional relationship of these functions with regard to cyberspace missions and support teams.

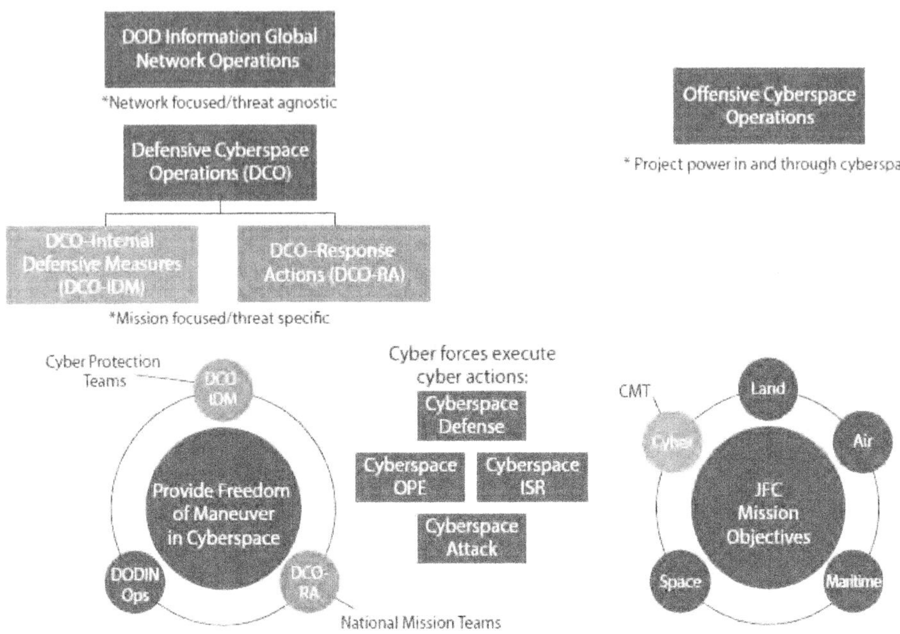

Figure 3. Cyberspace Operations Functional Relationships.[56]

Examining further details of these functions quickly leads to classified material that is inappropriate for this monograph. A capstone joint doctrine publication, *Joint Publication (JP) 3-12, Joint Cyberspace Operations*, was released in February 2013 for cyberspace operations for those readers with appropriate clearance and need to know. The unclassified synopsis states that the publication seeks to address "the uniqueness of military operations in cyberspace, clarify cyberspace operations-related command and operational interrelationships, and incorporate operational lessons learned."[57]

EVOLUTION OF ARMY CYBERSPACE OPERATIONS

Having examined the evolution of joint cyberspace operations, this section focuses on parallel evolutionary efforts in Army cyberspace operations toward the establishment of Army Cyber Command. It examines initial operations of the command under the leadership of Lieutenant General Rhett Hernandez as well as its current operations led by Lieutenant General Edward Cardon. This includes a brief review of recent efforts to establish Fort Gordon, Georgia as the center of gravity for Army cyberspace activities.

The Founding of Army Cyberspace Operations.

Just a few years before the formation of JTF-CND, the Army was making organizational changes to begin consolidating the operational of information systems. Since May 1984, the U.S. Army Information Systems Command (ISC) provided the service-wide management of five information disciplines: communications; automation; records management; printing and publishing; and visual information. Based on the perceived need for better control over regional communication and computer systems by Army major commands and theater commanders, ISC was disbanded, and the Army Signal Command created in September 1996. During the next 6 years, the command focused on strategic signal support to Army combat units worldwide. However, these units were equipped and resourced at the major command or theater level with little coordination. Thus, the Army-wide information system became increasingly nonstandard in their equipment and protocols at a time when threats to the system were growing more complex and widespread.[58]

To address these issues, the U.S. Army Network Enterprise Technology Command/9th Army Signal Command (NETCOM/9th ASC) was established in August 2002. Its mission was to "operate, manage, and defend the Army's 'Infostructure' at the enterprise level" to provide "Command, Control, Communications, Computers, and Information Technology common user services and signal warfighting forces in support of the Army, its Army service Component Commanders, and the Combatant Commanders." This included operation and defense of the Army's portion of the GIG.[59]

The USSTRATCOM 2005 CONOPS for GIG NetOps identified the Commander, U.S. Army Space and Missile Defense Command (USASMDC)/Army Forces Strategic Command (ARSTRAT) as the Army service component to JTF-GNO.[60] The Army NetOps structure had three tiers: (1) the central command element of the Army Network Operations and Security Center (ANOSC), referred to in the CONOPS as the Service Global Network Operations and Security Center (SGNOSC); (2) the combatant command support elements of the Theater Network Operations and Security Centers, referred to in the CONOPS as the Service Theater Network Operations and Security Centers; and support elements within theater of the Regional Network Operations and Security Centers.[61] Figure 4 depicts how the Army implemented this three-tiered structure across the five geographic combatant commands. The ANOSC[62] (or SGNOSC) at Fort Belvoir, VA, provided "decisionmakers a comprehensive, integrated, near real-time, situational awareness, [and] operational reporting capability" as well as "worldwide operational and technical support to the Land-WarNet across the tactical and strategic levels."[63]

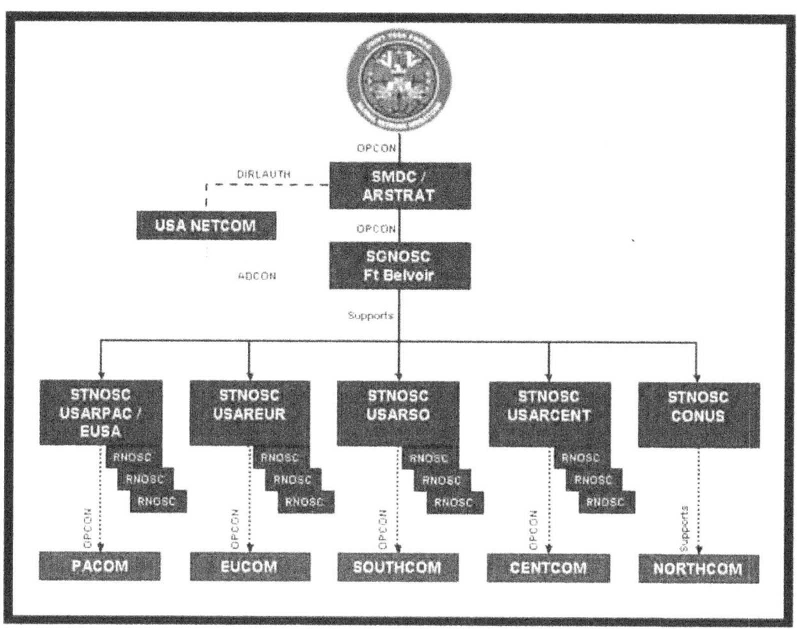

Figure 4. U.S. Army NetOps Forces (Circa 2005).[64]

In October 2006, the army reinforced the NETCOM/9th ASC mission and redesignated it as the U.S. Army Network Enterprise Technology Command/9th Signal Command (Army) (NETCOM/9th SC (A)). Its mission was clarified to formally include network-centric operations in context of the LandWarNet by executing:

> globally based and expeditionary communications capabilities to enable joint and combined battle command, leveraging the information grid to ensure extension and reachback capabilities to the warfighter.

It was to accomplish this "through globally postured theater signal commands, brigades, and regional information managers."[65]

Perhaps a good example of warfighter support facilitated by NetOps using the GIG is that of friendly force tracking (FFT). Originally called blue force tracking, the initial aim of the program was for U.S. Space Command to use national technical means "to provide a beyond line-of-sight, low probability of detection and interception, precise location of Special Operations Forces elements."[66] When U.S. Space Command merged with U.S. Strategic Command in 2002, the FFT mission operational control transitioned to USASMDC/ARSTRAT. In December 2008, the USSTRATCOM FFT mission was refined and USASMDC/ARSTRAT was given responsibility "to provide FFT data services on a continuous basis to combatant commands" and interagency and coalition users (when directed) as well as "to provide a combat development capability integrating FFT data into current and planned architectures for use on the appropriate Common Operating Picture."[67] The system has now become so integrated into joint operations that it may be taken for granted. Its continued success depends on coordinated NetOps support to generate, collect, process, disseminate, and display joint FFT information to warfighters worldwide.[68]

The 2009 version of the U.S. Army Posture Statement contained a summary of the Army's evolving cyber operations, which included descriptions of the NETCOM/9th SC defensive cyberspace focus of NetOps as well as the Army Intelligence and Security Command (INSCOM) offensive cyberspace focus of network warfare. By this time, Army cyberspace operations had been:

> integrated throughout Service and Joint Force structure, from strategic levels such as the Defense Information Service Agency, Joint Task Force-GNO, NSA, and Joint Functional Component Command-Network Warfare down to the Brigade Combat Team (BCT) level.

This included forward-based forces within theater signal commands, military intelligence brigades, and planning elements.[69]

Initial Army Cyber Command Operations.

In May 2009, the Army established a cyberspace task force to examine how to organize the service's cyberspace assets to support the anticipated establishment of a sub-unified command in USSTRATCOM dedicated to cyberspace operations. Specifically, the task force would synchronize the cyberspace-related activities of the Army Staff Intelligence/G-2, Operations/G-3, and Chief Information Officer/G-6. More importantly, it would examine if existing organizations (i.e., NET-COM, INSCOM, or USASMDC/ARSTRAT) could best provide the headquarters functions to direct the Army's existing cyberspace operation capabilities, or if a new command should be established. When Defense Secretary Gates issued his June 2009 memorandum to establish USCYBERCOM, the Army opted to retain USASMDC/ARSTRAT as the interim choice for U.S. Army Forces Cyber Command (ARFORCYBER).[70] At that time, the organization of Army cyberspace forces was largely the same as it had been described in the 2005 USSTRATCOM CONOPS, with a central command element and Theater Network Operations and Security Centers (TNOSCs) as well as Army Computer Emergency Response Teams (ACERTs). The Army Global Network Operations and Security Center (AGNOSC) remained essential to warfighting as "the Army's global eyes and ears in cyberspace . . . actively defending the Army's operational and generating force information capabilities from a continuously evolving, adaptive enemy." Also, TNOSCs continued their mission to "direct the operations, management and defense of the Army's portion of the link to the GIG."[71]

In February 2010, based on "the increasing global scope of the cyberspace mission," the Army chief of staff approved the establishment of a separate command for ARFORCYBER.[72] In June 2010, it was announced that Major General Rhett A. Hernandez would be the new ARFORCYBER commander with the task of achieving Army Cyber Command full operational capability by October 2010. While the roles of NETCOM/9th SC (A) and INSCOM remained largely unchanged, a new nerve center for Army cyberspace operations was created: the Army Cyber Operations and Integration Center (ACOIC).[73] With functions similar to those of the previous AGNOSC, the ACOIC was designed not only to provide Army forces with "clear, concise, and timely direction to execute full spectrum operations in cyberspace" but also to coordinate Army cyberspace operations and "to share information with other Army commands, our counterparts in the other services, and the U.S. Cyberspace Joint Operations Center." To facilitate this integration, some ACOIC personnel were physically embedded with the USCYBERCOM joint staff.[74]

As the organization charts were being redrawn for ongoing Army cyberspace operations, the Army Training and Doctrine Command (TRADOC) began a "Cyberspace/Electromagnetic Contest" capabilities based assessment in February 2010.[75] TRADOC also published the "Cyber Operations Concept Capability Plan 2016-2028" in February 2010 as the:

> first step in developing a common understanding of how technological advancements transform the operational environment, how leaders must think about

cyberspace operations, how they should integrate
their overall operations, and which capabilities are
needed.[76]

The report assessed that "the Army's current vo-
cabulary, including terms such as computer network
operations (CNO), electronic warfare (EW), and information operations (IO) will become increasingly inad-
equate."[77] It posited three interrelated dimensions of
full spectrum operations built upon these elements:
one of "psychological contest of wills;" a second of
"strategic engagement;" and the third dimension of
"the cyber-electromagnetic contest" — the focus of the
plan.[78] Arguing that cyberspace operations (Cyber
Ops) was more than the CNO and NetOps, the plan in-
troduced "four components for CyberOps: CyberSA,
CyNetOps, CyberWar, and CyberSpt, with CyberWar
and CyNetOps being the primary operational components."[79] The plan went on to develop an initial matrix
of required capabilities for each element in the areas of
doctrine, organizations, training, materiel, leadership
and education, personnel, and facilities.[80]

As planned, Army Cyber Command was established on October 1, 2010,[81] with a split-cased scheme
that had its headquarters at Fort Belvoir, and select
staff elements located with or near USCYBERCOM at
Fort Meade, MD.[82] Its mission was threefold: to lead
the planning and implementation of Army NetOps
and defense of Army networks; when directed, to
conduct cyberspace operations to ensure freedom of
action in cyberspace and to deny the same to adversaries; and to report, assess, and mitigate Army cyberspace incidents.[83]

Over the next year, several modifications were implemented to the initial U.S. Army Cyber command
(ARCYBER) organizations. In February 2011, Sec-

retary of the Army John M. McHugh issued a directive that the Army IO mission transfer to ARCYBER. Along with this new mission, ARCYBER received operational control over the 1st Information Operations Command (Land), which included IO support to warfighters using deployable teams that could leverage reach-back planning and analysis as well as synchronize and conduct CNO tasks. [84] In October 2011, the 780th Military Intelligence Brigade became ARCYBER's cyber brigade to serve as the command's "operational arm for full-spectrum cyberspace operations."[85] As such, the brigade was "organized to support USCC [USCYBERCOM] and combatant command cyberspace operations" as well as to conduct "signals intelligence and computer network operations, and enables Dynamic Computer Network Defense of Army and DoD networks."[86] ARCYBER also established the Army Cyberspace Proponent Office "to define the Army's future cyberspace force; design its organizations; establish the requirements to build it (both technological and human); and to develop the overarching cyberspace doctrine and operational constructs."[87] The command relationships resulting from these first-year changes are depicted in Figure 5.

During the first year of operation, ARCYBER did much to advance Army cyberspace operations along three lines of effort: operationalizing cyberspace; growing Army cyber capacity and capabilities; and recruiting, developing, and retaining Army cyber professionals. At a public conference in August 2011, Lieutenant General Hernandez discussed nine major accomplishments for the year that highlighted progress in the operationalization and unity of effort within the command.

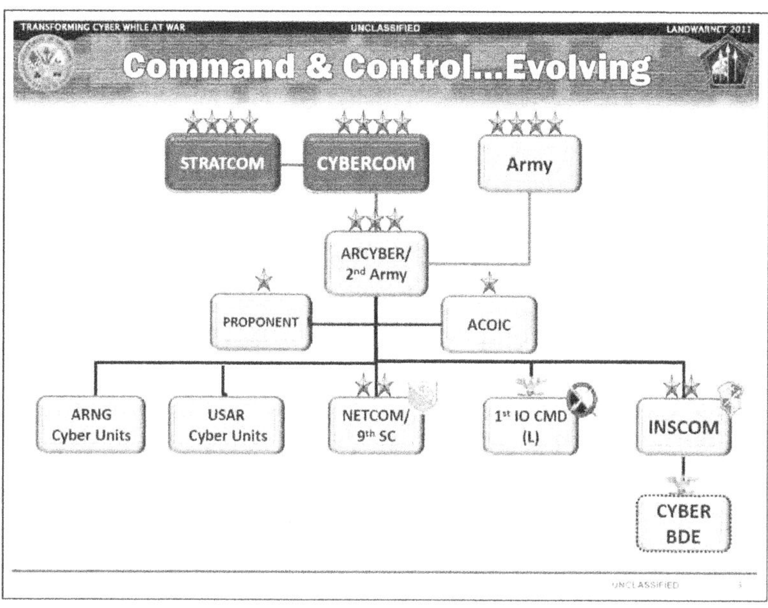

Figure 5. U.S. Army Cyber Command/Second Army (Circa 2011).[88]

Although these were significant steps forward, there still remained considerable work to achieve the commander's vision "to effectively defend our networks and deter and oppose our adversaries" as well as "to enable cyberspace activities under various authorities to work in concert with each other to more effectively support cyber operations."[89] Fundamental first steps in achieving these goals include improving our ability to see and understand our networks better. We will do this by collapsing our networks from a disparate, loose federation into one Army enterprise network. This will enable us to establish centralized control of our networks and give us more complete, integrated visibility into them. Having accomplished this, we can then establish an active defense in depth across the network.

Current Army Cyberspace Operations.

Looking toward the future, the 2012 *Army Posture Statement* identified three essential cyberspace elements to fulfill the needs of the dynamic information environment of 2020: a cyberspace enterprise; a "combined arms" cyberspace force; and integration, planning, and synchronization of cyberspace effects.[90] To fully incorporate these cyberspace elements into full spectrum operations, three cyberspace imperatives were set forth in the areas of personnel, cross-domain operations, and integrated operations. The personnel focus is to pursue "the development of Cyberspace Warriors and cyberspace formations to gain physical, temporal, and psychological advantages over an enemy will enable freedom of movement in, from, and through cyberspace."[91] The second imperative seeks to make cyberspace operations "routine and pervasive" given that "the Army will embrace cross-domain synergy between land and cyberspace. Cyberspace operations will be a critical part of 'How the Army Fights'."[92] The third imperative is probably the most challenging since it deals with several evolving mission areas: "Army Cyber will integrate and synchronize cyberspace operations with electronic warfare, electromagnetic spectrum operations, information operations, and space operations to achieve commander's objectives to ensure mission command."[93]

ARCYBER continued to evolve with efforts to address capability gaps identified in TRADOC's *Cyber/Electromagnetic Capability Based Assessment*. These included:

increase our [ARCYBER] World Class Cyber Opposition Force (WCCO) capacity to provide realistic, challenging cyberspace training in the conduct of Unified Land Operations to exercises, Home Station Training, and Combat Training Centers; increase our capability to conduct active defense of Army Networks through "Hunt Teams" that can find, fix, and mitigate currently un-detected malicious actors already inside the DoD infrastructure; provide capability to integrate cyberspace operations into Regional Army Land operations to support commanders' tactical and operational cyber planning and integration; increase intelligence personnel to support Army Cyber Command's operations Center, and improve our capability for rapid development of network defense tools; increase capacity to conduct our ability to conduct force modernization for cyberspace operations by developing requirements and solutions.[94]

In addition to these areas, ARCYBER also made progress in building relationships with allies and partner nations through participation in operational planning and Theater Security Cooperation effort with combatant commands.

In September 2013, ARCYBER/2nd Army welcomed its second commander, Lieutenant General Edward C. Cardon, who continued to build on the foundation created by Lieutenant General Hernandez. In his initial assessment of the command, Lieutenant General Cardon identified the three greatest continuing challenges as "building cyber capability and capacity; transitioning to a more defensible platform; and gaining situational awareness in cyberspace."[95]

In March 2014, the Army affirmed its commitment to unity of effort in cyberspace operations and refined the command relationships: making ARCYBER an Army Force Component Headquarters; designating

2nd Army as a direct reporting unit; and assigning NETCOM/9th SC (A) to 2nd Army, with Commander, NETCOM dual-hatted as the Deputy Commanding General, 2nd Army.[96] Figure 6 depicts the command relationship of this time frame.

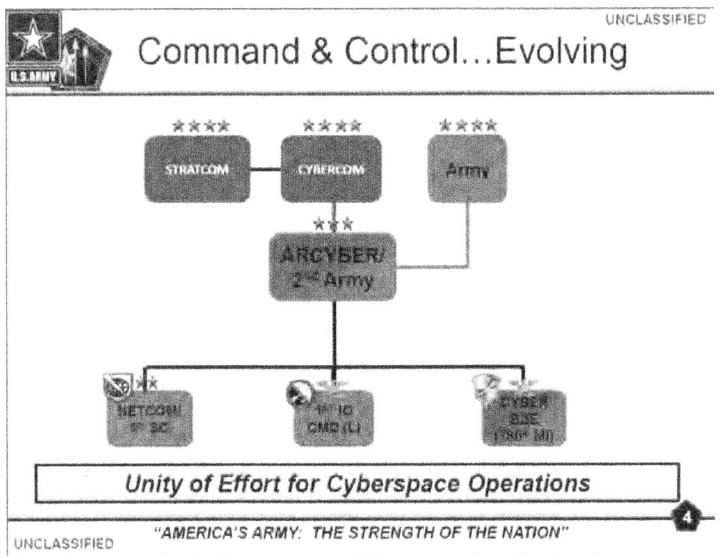

Figure 6. U.S. Army Cyber Command/Second Army (Circa 2014).[97]

After leading the command for 6 months, Lieutenant General Cardon offered additional refinements into these challenge areas, focusing on limitations of existing information architectures and cyber training as well as more strategic issues of risk assessment and authorities to match operating concepts. At the operational level, he discussed cyberspace operations in terms of maneuver on "cyber terrain" where one could replace traditional maps with "roads as [information] transport—fiber, satellite links, wireless. Think of the intersections as routers and switches, and think of the buildings as endpoints or people with mobile devices."[98] In such a schema, ARCYBER needs to recognize "there's a real nexus between land, cyber, and the human domains." At the strategic level, he noted that "cyber's a domain and it must be integrated with other domains to provide options to the National Command Authority."[99]

To help address these myriad tasks, ARCYBER is applying the total force concept to current Army cyberspace operations. For example, the 1st IO Command includes four Reserve Component Theater IO Groups with deployable capability that "provides IO and cyberspace planning, analysis and technical reach back; and offers specialized IO and cyberspace training to assist the warfighter in garrison, during exercises, or in conflict."[100]

Army National Guard (ARNG) units also play important cyberspace roles that may leverage technical experience from their civilian jobs. The Guard's 2015 Posture Statement summarizes some of the advantages this arrangement offers, to include unique legal authorities, knowledge of local critical infrastructure, and experience from work with commercial IT companies.[101] A specific application of this concept was initiated on June 5, 2014 when a memorandum of understanding was signed between ARCYBER/2nd Army and the ARNG to have the 1636th Cyber Protection Team serve in active Title 10 status in support of ARCYBER/2nd Army. The unit may be called upon to conduct any of the following missions:

defensive cyberspace operations, cyber command readiness inspections, vulnerability assessments, cyber operational forces support to emulate threats, critical infrastructure assessments, theater security cooperation and Federal Emergency Management Agency support.[102]

Probably the biggest change on the horizon for ARCYBER is the pending move of its headquarters to Fort Gordon, GA. The Army assessed this as the best option to address the need for additional space once the command outgrew its facilities at Fort Meade. In theory, moving to Fort Gordon is the least costly alternative. Also, the collocation of the Army's operational cyber headquarters with the Army's Joint Force Headquarters-Cyber and NSA-Georgia will require 150 fewer personnel.[103]

Part of the consolidation of Army cyber forces at Fort Gordon is the establishment of the Army Cyber Center of Excellence (CoE) there with goals of "aligning Army cyber proponency within TRADOC, creating institutional unity and a focal point for cyber doctrine and capabilities development, training, and innovation."[104] In fact, on March 28, 2014, the U.S. Army Signal CoE became the Army Cyber CoE with the initial fusion of various elements of cyber, signal, and electronic warfare training completed by October 2014 and full operating capability achieved by October 2015.[105] The new CoE is now responsible for the development of Army signal and cyber doctrine and is currently working to produce *Field Manual (FM) 3-12, Cyberspace Operations*, which will provide "tactics and procedures for the coordination and integration of cyberspace operations in support of unified land operations."[106]

The most significant current Army doctrine regarding cyberspace is FM 3-38, *Cyber Electromagnetic Activities* (CEMA), first published in February 2014. It provides "an overview of principles, tactics, and procedures on Army integration of CEMA as part of unified land operations." Further, it describes how Army "CEMA are implemented via the integration and synchronization of cyberspace operations, electronic warfare (EW), and spectrum management operations (SMO)."[107] Focusing on Chapter 3 of FM 3-38, the depiction of the doctrinal concept of cyberspace operations as three interdependent functions (see Figure 7) is consistent with terminology of USCYBERCOM.[108] While a worthy topic, the detailed analysis of the CEMA concept depicted in FM 3-38 is beyond the scope of this monograph.

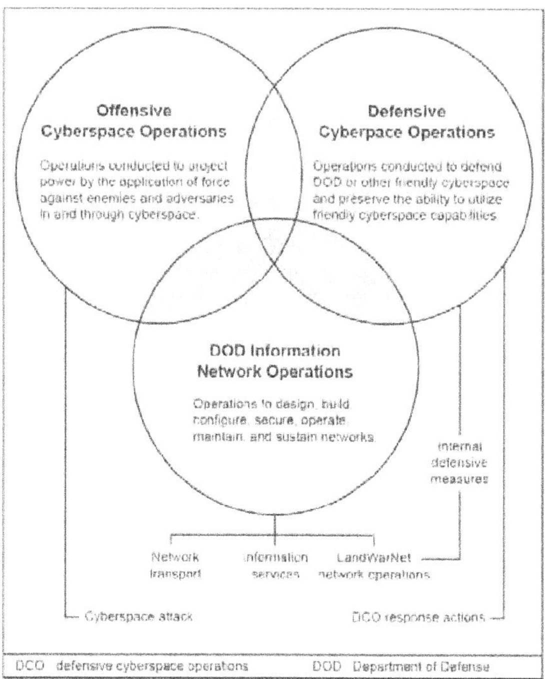

Figure 7. U.S. Army Cyberspace Operations Functions.[109]

Following the model of the Quadrennial Defense Review (QDR) and USCYBERCOM, ARCYBER implements its mission across four team structures: (1) Joint Force Headquarter-Cyber to provide operational and tactical planning support to Combatant Commands; (2) Cyber National Mission Force to defend the nation by seeing adversary activity, blocking attacks and maneuvering to defeat them; (3) Cyber Protection Force to defend DODIN and, when authorized, other infrastructure; and (4) Cyber Combat Mission Force to conduct military cyber operations in support of combatant commanders.[110] Figure 8 depicts how the goal of operationalizing cyber is achieved by combining these teams with the organization shown in Figure 8 and overlaying them across the ARCYBER mission areas.

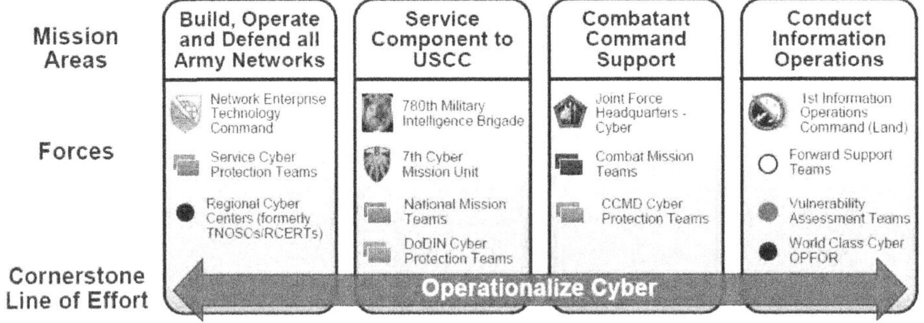

Figure 8. U.S. Army Cyberspace Operations Spectrum.[111]

A recent example of the continuing evolution of Army cyber forces to support these team structures is the 7th Signal Command (Theater) efforts to establish a new Cyber Mission Unit (Provisional) that will focus on defensive operations for Army networks. The new unit will form Cyber Protection Teams to "conduct global cyberspace operations to deter, disrupt, and help defeat the nation's adversaries in cyberspace. They will rapidly evaluate, and act proactively and reactively to dynamic cyber situations."[112]

CYBERSPACE OPERATIONS IN A GLOBAL CONTEXT[113]

Thus far, this monograph has addressed how cyberspace forces are currently being integrated across the full spectrum of traditional domain-based military operations. But is this approach sufficient to address the full scope of cyberspace operations now and into the future? This section takes a more theoretical slant to addressing this question as it examines an international environment of multiple actors interacting with varying degrees of tension. In such a global situation, cyberspace operations seeking to produce certain effects must also be examined for their potential to cause escalation of activities; possibly even up to the point of existential threat.

When the stakes become this high, then the topic of national deterrence comes into play. Indeed, one of the principles to guide development of the Joint Force of 2020 is to "include a renewed emphasis on the need for a globally networked approach to deterrence and warfare."[114] Admiral Rogers during his congressional confirmation hearing for the position of CDRUSCYBERCOM noted that "cyber warfare is a complex and evolving discipline, and the subject of deterrence is drawing increasing attention at all levels of government and the Interagency, and in our discussions with our international partners."[115]

A thorough examination of the topic of how all cyberspace operations influence, and are influenced by, global deterrence consideration may require several volumes of work. Instead, this monograph will introduce a methodology—the modified Herman Kahn Escalation Ladder—and use it to analyze the specific case of active cyber defense (ACD) operations. Readers may then modify and apply the analysis framework for their own needs. For our purpose, ACD is a concept that is currently embodied in the terms cyber defense in depth or DCO-RA.[116] The effective use of ACD as an instrument of national policy is not an isolated process with defined boundaries. Rather, it involves intertwined processes that transpire within a dynamic international environment. Ideally, such defenses will deter potential aggressors and work to defeat any who are not deterred. This section explores how ACD may integrate with traditional military operations across the spectrum of international conflict as well as how such defenses influence national responses related to deterrence and escalation.

A key aspect in addressing this issue is to explore such activities in the realm of existential threat, which traditionally is limited to nuclear warfare. Proper deterrence at this level can serve as an essential element of an overall risk reduction strategy to keep inevitable and unpreventable minor cyber incidents from escalating.[117] Thus, let us examine defensive and offensive cyber capabilities in the context of an expanded model for strategic deterrence that embraces and expands traditional nuclear deterrence. This approach reflects a more realistic international environment where major cyber attacks are not considered to be isolated events, but rather as one instrument among many aimed at achieving strategic goals.[118]

Kahn Model of Escalation and Deterrence.

Current U.S. military doctrine defines deterrence as "the prevention of action by the existence of a credible threat of unacceptable counteraction and/or belief that the cost of action outweighs the perceived benefits," but interestingly, the definition for escalation has been removed.[119] This change appropriately reflects the doctrine's focus on theater-level military operations using a six-phase model with a second phase of "Deter." The context for strategic deterrence focuses on influencing the decisionmaking of potential adversaries not to take actions that threaten vital interests. This is achieved through credible threat of action in three ways: denying them benefits; imposing costs; and encouraging constraint.[120] Implicit in this paradigm is the credibility to raise the stakes—escalate the conflict—to a point that is not acceptable by the adversary.

A famous model developed during the Cold War was Kahn's Escalation Ladder that he described as "a methodological device that provides a convenient list of the many options facing the strategists in a two-sided confrontation."[121] He illustrated his metaphor as a ladder with 44 "rungs" grouped into 7 larger crises regions of increasing intensity separated by distinct threshold events. His concept is useful to view the changes in conflict based on the interplay between the political, diplomatic, and military issues surrounding the conflict and the level of violence and provocation at which it occurs.[122] Although created in a different era of conflict, the Kahn ladder can be evolved and expanded to strategic warfare that includes other weapons in the deterrence force mix, such as global conventional strike and offensive cyber operations.[123]

The goal is not to replace nuclear forces, but rather, to develop a more holistic integration of strategic forces.

Simplified Escalation Ladder.

To examine a more integrated deterrence metaphor, let us first simplify the Kahn ladder by limiting it to the seven major crisis regions and their thresholds. In the original model, the Bizarre Crises region included five rungs that depict the initiation of actions related to limited nuclear warfare in various forms.[124] Let us divide these regions at the level of Bizarre Crises into a lower half group that encompasses conflict at the theatre/regional level and an upper half group that addresses existential conflict (see Figure 9).

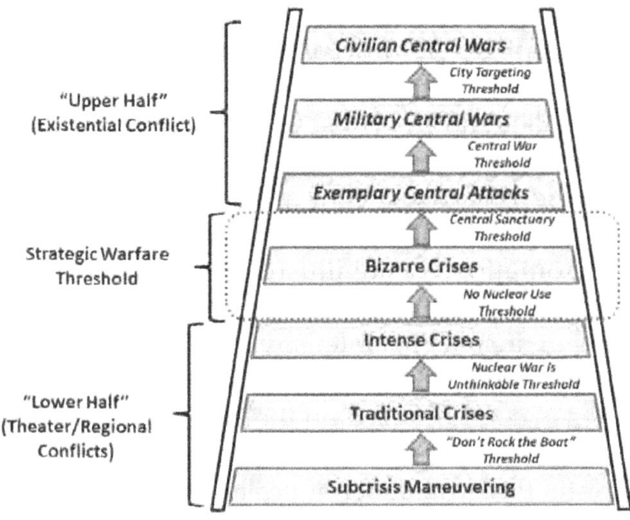

Figure 9. Modified Kahn Escalation Ladder.[125]

The lower half of the simplified ladder starts with Subcrises Maneuvering, which consists of political, economic, and diplomatic gestures, as well as formal declarations, to demonstrate resolve. When military forces come into play, the activity crosses the threshold to Traditional Crises. In this region, activity increases progressively from shows of force and mobilization, through harassing acts of violence, and up to dramatic confrontations. When military forces become the main focus of conflict, the activity crosses the threshold to Intense Crises, and the view of nuclear stockpiles change from hypothetical to realistic threats. In this region, diplomatic measures support coercion using provocative acts such as ultimatums, embargos, and blockades. Conventional conflict increases in its scope and intensity toward the formal declaration of war and movement closer to the incorporation of nuclear weapons.[126]

The upper half of the simplified escalation ladder deals with conflict that has escalated to the point of potential existential threat of nuclear attack. It begins with Exemplary Central Attacks where nuclear weapons are used in a restrained manner against specific military, infrastructure, or population targets. As activities progress through the ladder rungs, reciprocal reprisals occur. When military forces become the main focus of nuclear weapons, the activity crosses the threshold into Military Central Wars. In this region, military commanders have access to all the resources of the nation as well as nuclear weapons, but they use tactics that limit collateral damage to an opponent's civilians. Its rungs progress from targeting specific property and forces in equal responses, to constrained force-reduction attacks, then to increasingly intensive counterforce strikes using nuclear weapons. When these counterforce strikes exceed any attempt to spare civilians, then the activity crosses the final threshold into Civilian Central War. This is the region of nightmarish nuclear exchanges that devolve from "city-trading" attacks of resolve, to purposeful destruction of the enemy's society, and ultimately to the insensate launch of all weapons without regard to consequences.[127]

Movement Along the Ladder.

Kahn designed his ladder metaphor to examine the interrelations between two sets of elements surrounding a given escalation situation—those specific to the region of the present conditions and those related to the dynamics of moving on the ladder. He envisioned the ladder to model two-sided escalation (usually the United States and the Union of Soviet Socialist Republics) that met certain conditions related to: commitment of resources; value placed on victory; interest in systems bargaining to preserve precedents; motivations and strategies for escalation; desire to appear to be following accepted norms; and danger and avoidance of upper levels of escalation. He divided national conduct related to movement on the ladder into five categories: contractual (*quid pro quo*); coercive (stick versus carrot); agonistic (prescriptive rules); stylistic (accepted norms), and familial (positive cultural aspects). As one might expect, activities in these categories would reflect the use of all elements of national power (political, economic, information), and thus Kahn asserted that "mere military superiority will not necessarily assure 'escalation dominance'."[128]

Admittedly, the paradigm is not perfect. The movements reflecting escalation are not necessarily sequential, symmetric, or reversible. Also, the ladder is not very useful at illustrating the effects of multiple simultaneous moves. Any analysis should also recognize that an adversary will also have a ladder (implicit or explicit) that is likely different in its placement and perception of conditions. It also assumes the interactions involve rational players in a model that often fails to fully embrace ambiguity and uncertainty related to acceptable alternatives and long-term stability.[129] Regardless, the simplified ladder offers a reasonable framework to examine an integrated strategy of deterrence.

Examining Escalation and Deterrence.

With the foundation of the simplified escalation ladder, let us apply it to a broader view of strategic warfare that includes conventional global strike and cyber offensive forces in addition to nuclear forces to provide deterrence across domains. Once this is codified, we can then examine the roles of ACD in the paradigm. To be clear, this is not an examination of a cyber escalation ladder developed by Dunn Cavelty.[130] Nor is it akin to analysis by Martin Libicki that downplays valuable lessons from the Cold War and considers "cyber escalation" largely in isolation.[131] Rather, this analysis addresses a more evolutionary and holistic view of modern deterrence and warfare with a scope emphasizing various forms of the military instrument of power. For the scope of this monograph, examples of national policies and doctrines will be drawn from those of the United States.

Types of Warfare and Factors.

Conflict in the lower half of the simplified ladder involves the evolving forms of conventional and irregular warfare at the theater/regional level. Military forces are organized, trained, equipped, and employed in traditional domains, but they also adopt activities in the cyberspace realm as an integral part of joint operations.[132] The U.S. concept of globally integrated operations provides guidance and details for a force that by 2020 can "quickly combine capabilities with itself and mission partners across domains, echelons, geographic boundaries, and organizational affiliations."[133] These would incorporate existing teams from USCYBERCOM that "operate and defend the networks that support military operations worldwide" as well as "support combatant commanders as they execute military missions."[134] Conflicts would strive to protect national interests and achieve stability in the given region with approaches that adhere to internationally acceptable norms. Kinetic attacks would emphasize precision of targeting and delivery as well as predictable results that are appropriately limited in first order and collateral effects.

In the upper half of the simplified model, conflict has escalated to the point where vital national interests are threatened, potentially to the degree of existential vulnerability. To deter or confront such threats, consider a military force structure that adds protected conventional strategic strike and offensive cyber capabilities to traditional nuclear forces delivered by aircraft or long-range missiles. This concept developed by the U.S. Defense Science Board maintains the need for cyber defense of an overarching nuclear capability as well as a portion of conventional global strike forces that are segmented from similar lower half forces to receive enhanced cyber survivability measures.[135] Akin to the original Kahn model, attacks will intensify to counterforce targets and then broaden to civilian infrastructure toward a worst case of being totally indiscriminate. Conflicts at these degrees of escalation are likely to operate outside of accepted international norms, or perhaps even in ways where no norms exist. Weapon delivery precision, effect predictability, and collateral damage avoidance become more difficult due to the increased intensity of operations as well as less important when compared to the increasing national stakes.

The strategic war threshold between the lower and upper escalation areas is no longer limited to the use of nuclear weapons, and, in fact, it is highly unlikely that any limited nuclear exchange would occur. Rather, this becomes the region where limited offensive cyber or conventional global strike may begin against vital targets found in the upper half. Such strikes could have effects beyond the accepted proportionality and perfidy of those in the limited conflict, whether by design or by accident. Thus, it is crucial for forces to be cautious in the use of such weapons to minimize unintended consequences that may cross into the upper half of the ladder.

Dynamics of Conflict.

In the lower half, Kahn notes there are three main ways to escalate a limited conflict: increase its intensity; widen the area; or compound the escalation by attacking other actors. He offers an analogy for this area's dynamics as being similar to those of a labor strike. In each case, it is assumed that both sides have serious issues to resolve, sometimes through threats of harm, but there is no real desire to do permanent or excessive damage. As with a labor strike, the conflict may require considerable give-and-take bargaining to ensure stability between the parties.[136]

In contrast, conflict in the upper half of the ladder can be likened to a game of "chicken," a contest of brinksmanship that creates a winner when the loser loses their nerve (such as driving two cars toward each other to see who will swerve to avoid a collision). Unfortunately, in the worst case, both parties are destroyed (no one swerves), and in the best case, the loser is humiliated, leaving little chance for compromise or face saving necessary for long-term stability.[137] Thus, a strategy of deterrence should include widely understood precedents and thresholds to be reliable for stability and controlled escalation that can prevent a game of chicken being played with nuclear weapons.

Roles of Active Cyber Defense.

As previously noted, the term ACD has no universal definition, but it is generally considered to include proactive measures that may extend beyond the particular network being defended. The roles of ACD and their relation to the dynamics of conflict and escalation can be illustrated as the ladder turned on its side as in Figure 10. In the lower half of conflict, the reality that there will always be minor cyber probing and attacks has been accepted and planning guidance now addresses resiliency for operating in a degraded network environment. For the U.S. military, the ACD is a "synchronized, real-time capability to discover, detect, analyze, and mitigate threats and vulnerabilities" which includes proactive operations "at network speed by using sensors, software, and intelligence to detect and stop malicious activity before it can affect DoD networks and systems."[138]

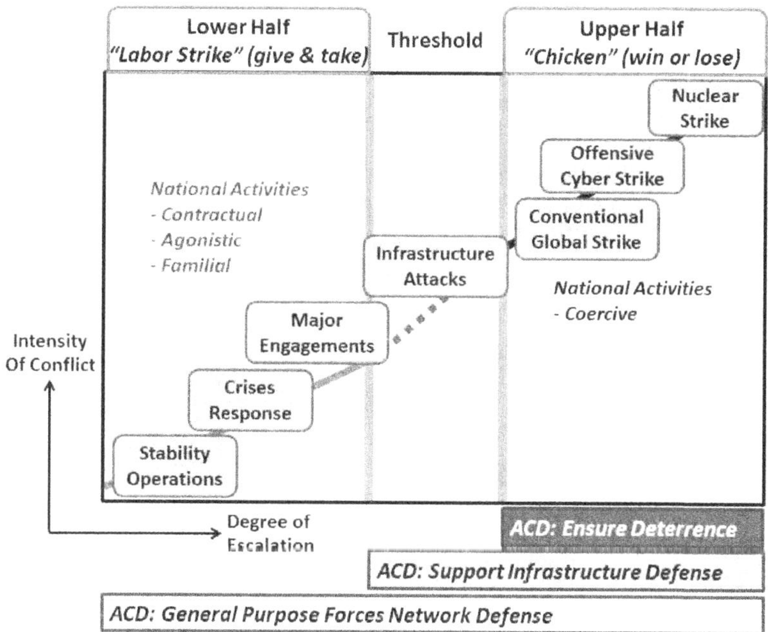

Figure 10. Relation of ACD to the Dynamics of Conflict and Escalation.

Ideally, ACD applications are limited to achieve the minimal effects necessary to defend the military network. This reflects several forms of national motivation; primarily contractual—working toward a reasonable cost/benefit balance—as well as agonistic—functioning along the lines of evolving rules of Internet governance. Motivations may also reflect familial norms, such as trying to preserve a free and open Internet. Stylistic motivations and actions may be a source of friction in limited conflict since they are often tied to national character and culture, which can vary greatly for cyberspace issues among the United States, Russia, China, North Korea, and others. Motivations of explicit coercion are not expected unless one is willing to accept possible escalatory consequences.

In military terms, any ACD actions that extend beyond blocking network access points would strive to be precise, proportional, and limited in scope. The focus would be to enhance joint operations of general purpose forces at the tactical and operation levels—mainly intelligence gathering and defenses that operate under decentralized authorities.[139] If kinetic attacks reach the level of armed conflict, then supporting cyber operations should also follow the tenets of the Law of Armed Conflict (e.g., necessity, distinction, proportionality).[140] As such confrontations occur in the future, systems bargaining among nations may lead to the development of formal and informal rules of engagement that add stability and reduce the chance for unintentional escalation. Certainly, nonstate actors can and do operate in cyberspace asymmetrically and outside of international norms, but that is beyond the scope of this discussion.

In the upper half of Figure 10, the goal is to prevent conflict from escalating to a game of chicken with nuclear arms. Of course, a strategy of deterrence requires the capabilities and resolve to conduct extreme violence in order to influence a potential adversary not to pursue such a course of action. If such forces are used, the concern for precision would focus on effectiveness with decreased concern for limiting collateral damage. Similarly, the criteria for distinction of purely military targets, especially in the cyber realm, may be relaxed in order to protect critical deterrent forces.

A prudent force structure in this case is to have separate ACD capabilities that are optimized to ensure the proper function of the deterrent forces — conventional strike, cyber offense, and nuclear strike. This approach also makes sense from a budget and resource perspective since the expense in adding additional protection, survival, and resilience measures would be confined to the critical portion of strategic ACD. Operations at this level would require "fires" authority that "should reside at the highest levels of government" with no decentralization.[141] This is consistent with traditional nuclear operations concept of execution direction being provided by a limited number of national command authorities. The national motivation leans heavily toward coercion after diplomatic efforts become increasingly strained and ineffective.[142]

Clearly, the threshold area is a critical transition from regionally limited conflict that largely conforms to international standards to a much riskier engagement that can escalate to existential stakes. In this area, kinetic activity has reached the levels of armed attack or perhaps armed conflict, and belligerent cyber activity has gone from minor probing and isolated intrusions to more complex and pervasive attacks. Criteria discussed in the *Tallinn Manual* can help assess its international legal implications,[143] but if the state-sponsored attacks begin against such targets as banks and power grids, the intensity and stakes move toward the upper half. While military ACD will still be operating at the tactical and operational levels, there needs to be additional measures of ACD extending to help protect against attacks on civilian and infrastructure targets. Chairman of the U.S. Joint Chiefs General Martin Dempsey recently noted about such cyber aggression: "It's not just an inconvenience, if we lost critical infrastructure on the east coast for a period of time, people's lives would be lost." The ACD required to protect cyber targets outside of military networks would be broader in scope and require interagency consultation, cooperation, and resources.[144] Potential ACD actions by citizens and private industry touch on many unresolved controversies that merit further discussion.

Table 2 summarizes the types of forces expected at each area of the simplified model; ACD is considered as a subset of cyber forces. Allied and coalition military forces would also be present at each level and the added complexity of their operations merits more detailed analysis beyond this monograph. Circumstances will dictate where activity begins along the escalation ladder; it need not begin at the lowest point. Any ensuing escalation need not be sequential or linear in its progression. Kahn offered several criteria to consider for measuring the degree of escalation possible in any particular time which in turn can indicate the scope of ACD required.[145] First, one must examine the current scale, scope, and intensity of violence of the conflict as well as the resolve (or recklessness) demonstrated. Next, one should assess if any actual damage has been done. What is the apparent closeness to war moving to the upper half of the ladder? Evaluating the stability of the conflict is important to determine the likelihood of eruptions or spikes in attacks that could fuel escalation. This would include evaluating what provocation has occurred and what precedents have been broken as well as what threats has been intended or perceived.

Escalation Ladder Area	Type of Military Forces		
	Conventional Forces	Cyber Forces	Nuclear Forces
Upper Half (Existential Conflict)	- Segmented from general forces - Precise Effects - Collateral damage more acceptable	- ACD focused on protecting deterrence - Triggers and activity authorized by highest national command - Cyber offense used	- Full alert for use - Aircraft & missile delivery - Weapons of last resort authorized by highest national command
Strategic Warfare Threshold	- Continued theater level conflict - Support of other agencies and departments	- Whole-of-government operations - ACD help support defense of national infrastructure	- Not used - Readiness increased
Lower Half (Theater / Regional Conflicts)	- General purpose forces in all domains - Precise delivery and effects - Minimize collateral damage	- ACD at network bounds - Limited ACD beyond network - Military command (delegated authority)	- Not used - Readiness maintained

Table 2. Use of Military Forces in Simplified Escalation Ladder Areas.

Active Cyber Defense and Deterrence.

Since an expanded deterrent capability with survivability enhanced by ACD measures plays an essential role in controlling conflict escalation, there is merit in a more detailed review of an implementation concept possible for U.S. forces. Figure 11 depicts a conceptual design for ACD interfaces supporting deterrence operations in the upper half of the escalation ladder. The ACD activities would operate in two modes: an automatic mode with triggers based on *a priori* criteria established and updated by command authorities and a manual mode that requires command authority direction for execution. Situational awareness is maintained through information provided by strategic intelligence sources as well as tactical and operational indications and warnings. Results from ACD actions—cyber battle damage assessment—are provided as feedback. Decisionmaking by national command authorities can be supported by artificial intelligence systems that can develop and assess courses of action, perhaps leveraging advanced "mindreading" designs that can rapidly perform modeling, simulation, and prediction reflecting fifth-order beliefs.[146]

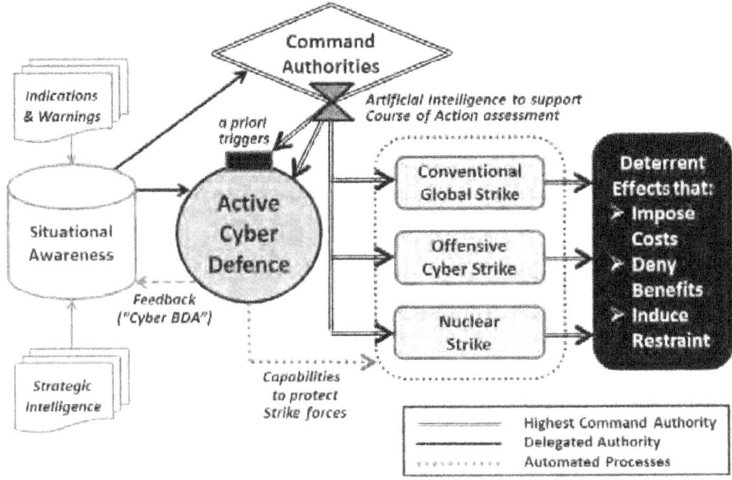

Figure 11. Details of ACD in Deterrence Operations.

The ACD system would provide continuous automated protection for the deterrence strike forces shown as well as the command and control systems of the command authorities. The first line of kinetic deterrent forces would be conventional global strike forces that are always segmented from general purpose forces—thus no dual purpose missions are allowed for these forces in limited conflict. These would be of sufficient quantity for anticipated threats, perhaps as few as 20 long-range aircraft plus long-range missiles. The ultimate deterrent remains nuclear forces, which would continue to be a mix of weapons delivered by aircraft and land- and sea-based ballistic missiles in numbers that reflect continuing arms reduction.[147]

The specific roles of offensive cyber strike forces are currently ambiguous and activities may overlap between ACD that assertively negates cyber attacks against deterrence forces and offensive cyber attacks for counterforce operations. The 2011 U.S. *International Strategy for Cyberspace* includes a declaratory statement that supports its inherent right to self-defense and deterrence: "When warranted, the United States will respond to hostile acts in cyberspace as we would to any other threat to our country." It goes on to state that such response may "use all necessary means—diplomatic, informational, military, and economic—as appropriate and consistent with international law."[148] Healey and Wilson examined cyber offensive actions and their approximate physical world equivalent and how existing executive and legislative provisions may apply to them.[149] A recent study by The Defence Academy of the UK cautions that "online weapons may be unreliable or uncertain in their effects" and that such weapons "coupled with an explicit policy of conventional military kinetic retaliation risks rapid escalation of real-world war."[150] Other respected theorists such as Colin Gray are more conservative in their assessments, offering that "cyber offense usually is likely to achieve some success," but that "the harm we suffer is most unlikely to be close to lethally damaging;" concluding that "it is clear enough today that the sky is not falling because of the cyber peril."[151] Clearly, the topic of integrating cyber offensive into strategic operations requires further extensive study.

Deterrence Effectiveness.

Perhaps some Cold War lessons learned can serve as a "litmus test" for an updated deterrence strategy incorporating ACD and cyber offence. Richard Kugler posits that U.S. nuclear deterrence worked because it was credible; it was conducted in the context of political dynamics; it denied the Soviet Union any favorable prospects from aggression; it favored development of flexible options; and it minimized the risk of unwanted escalation.[152] Incorporating ACD into deterrence improves credibility by enhancing deterrence force capabilities and survival. Also, having a declaratory statement from the country's executive in an official public document demonstrates resolve and legality. As Eric Jensen noted, "while this statement was controversial when made, there is no doubt of its legality."[153] The updated escalation ladder adds perspective on how to view ACD and other cyber support of operations not in isolation, but in the context of all elements of national power. Admittedly, this section has viewed these issues from the perspective of the United States, which implicitly includes mutual military commitments with allies; further discussion should examine this more explicitly. Having a three-pronged deterrence force protected by ACD strives to influence an adversary's decisionmaking by not only denying benefits, but also by imposing costs and inducing restraint. Implementing such a cross-domain framework "would contribute to more effective deterrence and crisis management."[154] By design, this cross-domain force provides national command authorities with flexible options that are beyond nuclear-only in case of extreme escalation.[155] In theory, while having more options below the nuclear level may reduce the chance of reaching the ultimate limit of war, there is no guarantee that it would minimize the risk of unwanted escalation below that threshold.

RECOMMENDATIONS

This monograph examines the past and present of joint and Army cyberspace military operations as well as how these operations may fit into the complex and dynamic sphere of international deterrence and escalation. To facilitate the best evolutionary path for future activities it recommends the following actions be considered.

Current Military Cyberspace Priorities.

The five command priorities set forth by General Alexander and carried forward by Admiral Rogers seem appropriate for the current evolution of US-CYBERCOM and progress on them continues at a steady pace. However, some of the successes in operationalizing cyberspace are hidden behind questionable classification decisions. Specifically, it is difficult to comprehend why the inaugural version of JP 3-12 was issued as a secret document instead of an unclassified document with a classified annex. This unnecessary occlusion of basic doctrinal tenets (such as those in FM 3-38) greatly hampers both U.S. and allied planners and military educators. This is particularly ironic when one considers that the former manifestation of JP 3-12 was as *Doctrine for Joint Nuclear Operations*, a document that was somehow kept unclassified. As cyberspace doctrinal information is incorporated in updates of capstone documents (e.g., JPs 3-0 and 5-0 [*Joint Operation Planning*]), the developers should consider adding a concise cyberspace annex that serves as a primer for cyberspace domain considerations. Military and national cyberspace activities writ large would benefit greatly if dedicated cyberspace theory development was promulgated that includes exploration beyond the domain definition of cyberspace. All of these recommendations could be supported by efforts at the Army's fledgling Cyber Center of Excellence.

Authorities.

Determining the appropriate authorities involved with decisionmaking and cyberspace operations, such as ACD actions, through the escalation ladder will continue to be a challenging and evolving issue. Military forces are developing doctrine and force structures to incorporate existing cyber related forces as well as newly defined positions. Ideally, these are tested, refined, and validated in exercise situations before full employment. However, as conflict escalates, so does the need to coordinate military operations with other powers of government as well as with allies and international governance bodies. Potential ACD actions by citizens and private industry and their impact on the conflict environment also have responsibility and legitimacy issues that cannot be ignored. At the high-stakes end of operations, one of the greatest challenges is determining ways of applying and updating the *a priori* authorities for ACD protecting deterrence forces. Jensen offers a detailed and nuanced assessment of legal issues related to cyber deterrence.[156]

Strategic Communication.

As work progresses toward better definition of cyberspace force roles based on context and dynamics of escalation framework, this must include strategic communication. These are planned and coordinated activities to provide the actions, images, and words necessary to help make the modified deterrence effective in the ways intended. Manzo notes that:

> cultural differences, contrasting strategic objectives, differing strengths and vulnerabilities can cause decisionmakers in the United States and other countries to reach different conclusions about proportionality and escalation.[157]

Efforts to overcome such differences could leverage studies like Melissa Hathaway's recent development of a Cyber Readiness Index, which examines the maturity and commitment for cybersecurity by 35 countries, including those that had formally established national strategies and competent authorities, mostly in nonmilitary areas.[158] Also, the publication of an unclassified version of JP 3-12 would contribute to the international understanding and commitment of U.S. cyberspace forces. All of these activities would support strategic engagement—the socio-political support for cyberspace operations—as the second dimension of full spectrum operations.[159]

Multi-Role Modeling.

Creating a realistic model for cyberspace force roles in escalation and deterrence requires a holistic consideration of environmental influences. As Ronald Deibert notes, "Securing cyberspace requires reinforcement of restraint on power, including checks and balances on governments, law enforcement and intelligence agencies."[160] The first dimension of full spectrum operations involves the psychological contest of wills.[161] The Kahn ladder was never envisioned for application beyond modeling interactions between two nations. To portray our multipolar world more accurately, models need to not only consider interactions between multiple nations, but also that the "policies to deter one type of adversary may differ from those needed to deter another adversary, with varying degrees of soft and hard rhetoric or of positive incentives and punishing responses."[162] The model should also include the dynamic of groups of nations, especially those in formal alliances such as the North Atlantic Treaty Organization (NATO). Finally, the activity of individuals and nonstate actors groups—some operating within accepted international norms, some not—can present asymmetric challenges and potential threats to the dealings amongst nations and thus should be included in the multi-role models.

Other Paradigms and Factors.

In addition to considering Cold War models such as the Kahn ladder, Sean Lawson also examined other metaphors as frameworks for analyzing cyberspace activities related to strategic deterrence. He posits there are similarities between insurgency or biological warfare and cyber crime and espionage.[163] Paradigms are needed to model cyber activity outside of designated military networks; these could help better define the threshold separating ACD that negates cyber attacks against deterrence forces from offensive cyber attacks for counterforce operations. Finally, the longer-term dynamics of de-escalation and counter-proliferation measures, such as potential arms control in cyberspace, introduce valuable methods for achieving and maintaining a more stable international environment in all domains. [164]

CONCLUDING REMARKS

Military cyberspace operations have been ongoing since before the advent of the Internet. Such operations have evolved significantly over the past 2 decades and are just now emerging into the realm of military operations in the traditional domains of land, sea, and air. To facilitate the operationalization of this new domain, education of the tenets of cyberspace must occur at the tactical, operational, and strategic levels of leadership. More importantly, the deliberate pursuit of understanding the full scope of cyberspace beyond that of a mere domain is essential for providing a theoretical foundation for current and future operations. Also in this regard, the development of such fundamental theory should look forward to embrace potentially radical manifestations of cyberspace in the future as well as looking back at its history.

The persistent increase of cyberspace activities in global events continues to make international dynamics more complex. The scope of context for such matters needs to consider not just other military efforts or even other instruments of national power, but how they are presented in an escalation framework and where they may be going. A modified Kahn escalation ladder is a useful metaphor to explore how cyberspace activities may integrate with traditional military operations across the spectrum of international conflict as well as how such defenses influence national responses related to deterrence and escalation. Expanding deterrence forces to include conventional strike and cyber offense can add capability and credibility as well as flexibility to course-of-action development available for national command authorities. Cyberspace operations such as automated cyber defense can support and enhance deterrence operations and limited conflict as well as help control escalation and reduce risk.

ENDNOTES

1. "Joint Task Force on Computer Network Defense Now Operational," Department of Defense News Release No. 658-98, Washington, DC: Department of Defense, December 30, 1998.

2. Greg Rattray, *Strategic Warfare in Cyberspace*, Cambridge, MA: The MIT Press, 2001. Chap. 5, "The United States and Strategic Information Warfare, 1991-1999: Confronting the Emergence of Another Form of Warfare" provides a thorough history of national security events leading up to the formation of JTF-CND.

3. Robert J. Lamb, "Joint Task Force for Computer Network Defense," *IA Newsletter*, Vol. 2, No. 3, Winter 1998/1999, pp. 3-4. The full mission statement was:

> Subject to the authority, direction, and control of the SECDEF, JTF-CND will, in conjunction with the unified commands, Services, and agencies be responsible for coordinating and directing the defense of DoD computer systems and computer networks. This mission includes the coordination of DoD defensive actions with non-DoD government agencies and appropriate private organizations.

4. "Joint Task Force on Computer Network Defense Now Operational," DoD News Release No. 658-98.

5. William C. Story, "Military Changes to the Unified Command Plan: Background and Issues for Congress," Congressional Research Service Report No. RL30245, Washington, DC: Congressional Research Service, June 21, 1999, p. 14.

6. Rudy de Leon, "Department of Defense Chief Information Officer Guidance and Policy Memorandum No. 6-8510 'Department of Defense Global Information Grid Information Assurance'," Washington, DC: Deputy Secretary of Defense, June 16, 2000, p. 8. Details of these missions areas were:
  - Coordinate and direct DoD-wide computer network defense operations to include:
    - Actions necessary to synchronize the defense of DoD computer systems and networks (e.g., network patches, firewall rules);
    - Actions necessary to stop a computer network attack (CNA) or computer network exploitation (CNE), limit damage from such activities, and coordinate the restoration of effective computer network service following a CNA or CNE;
  - Declare changes in INFOCON and issue INFOCONs in accordance with Chairman of the Joint Chiefs of Staff Memorandum CM-510-99, "Information Operations Condition (INFOCON)."

7. "National Defense Authorization Act for Fiscal Year 2002—H.R. and Oversight of Previously Authorized Programs before the Committee on Armed Services House of Representatives," Report HASC No. 107-5, 107th Cong., 1st Sess., Washington, DC: U.S. Congress, May 17, 2001, p. 17. In his testimony, the Honorable Linton Wells, II, Assistant Secretary Of Defense for Command, Control, Communications And Intelligence (Acting), and Department Of Defense Chief Information Officer, summed up the change from JTF-CND to JTF-CNO as:

We also need to coordinate cyber events better and responses. Solar Sunrise in February of 1998 found us with virtually no means to address the kind of problems we are facing. By the end of 1998, the Joint Task Force Computer Network

Defense had been stood up on an interim basis reaching full operational capability the following year. By October of 1999, the network attack mission had been passed to Space Command (SPACECOM). And in April of this year [2001], the Joint Task Force Computer Network Operations was put together. So there really has been a significant amount of progress in not a very long period of time.

8. *Ibid.*, pp. 23-26. Major General James Bryan went on to describe some of the Service and agency contributions:

The relationships upon which the JTF depends are its most important characterization. We have, as you can see, the Computer Emergency Response Teams of each of the four services and the Defense Information Systems Agency as the tactical components of our CND mission. We execute direction of the defense of the networks through these organizations. Without them, we would not be able to do our job.

9. Edward J. Drea *et al.*, *History of the Unified Command Plan*, Washington, DC: Joint History Office, 2013, pp. 86-87.

10. Jason Healey, ed., *A Fierce Domain: Conflict in Cyberspace, 1986 to 2012*, Washington, DC: Cyber Conflict Studies Association, 2013, pp. 65-66.

11. Harry D. Raduege, Jr., "Future Defense Department Cybersecurity Builds on the Past," *Signal*, Vol. 62, No. 6, February 2008, p. 120. Lieutenant General Raduege also noted that:

It was no accident that the secretary assigned the DISA director additional responsibility as the first JTF-GNO commander. DISA's extensive capabilities form a powerful platform in supporting emerging national-level and Defense Department cybersecurity requirements.

12. U.S. Strategic Command, *Joint Concept of Operations for Global Information Grid NetOps*, Offutt AFB, NE: U.S. Strategic Command, August 10, 2005, pp. 1-2. In this CONOPS, "NetOps is defined as the operational construct consisting of the essential tasks, Situational Awareness (SA), and C2 that CDRUSSTRAT-COM will use to operate and defend the GIG," and the GIG is defined in part as (p. 1):

Globally interconnected, end-to-end set of information capabilities, associated processes, and personnel for collecting, processing, storing, disseminating, and managing information on demand to warfighters, policy makers, and support personnel. The GIG includes all owned and leased communications and computing systems and services, software (including applications), data security services, and other associated services necessary to achieve Information Superiority.

13. *Ibid.*, p. 8.

14. *Ibid.*, pp. 13-14.

15. *Ibid.*, p. 27. Additional details of the theater-level cyberspace operations include:

The specific roles of the TNCC include monitoring of the GIG assets in their theater, determining operational impact of major degradations and outages, leading and directing responses to degradations and outages that affect joint operations, and directing GIG actions in support of changing operational priorities. The TNCC also responds to JTF-GNO direction when required to correct or mitigate a Global NetOps issue.

16. *Ibid.*, p. 35. Daily operations were facilitated as follows:

Additionally, the GNCC has DIRLAUTH [direct liaison authority] with the TNCCs. This authorization gives the GNCCs and TNCCs the ability to directly coordinate scheduled changes in the GIG or troubleshoot outages.

17. *Ibid.*, p. 36.

18. *Ibid.*, p. 18. Details of the JTF-GNO support included:

JTF-GNO directs the operation and defense of the GIG to assure timely and secure net-centric capabilities across strategic, operational, and tactical boundaries in support of DoD's full spectrum of warfighting, intelligence, and business domains.

The Commander, JTF-GNO (Cdr, JTF-GNO) will exercise Operational Control (OPCON) of the GIG for Global NetOps issues. Under the authority of CDRUSSTRACOM, JTF-GNO issues the orders and directives necessary to maintain the assured service of the GIG, ensuring that the President, SECDEF, CC/S/As [Combatant Commands, Services, and Agencies] can accomplish their missions. The CC/S/As execute JTF-GNO's directives within their respective areas and report compliance.

19. *Ibid.*, p. 2.

20. *Department of Defense NetOps Strategic Vision*, Washington, DC: DoD Chief Information Officer, December 2008, p. 4.

21. *Ibid.*, pp. 7-11.

22. Department of Defense Instruction 8410.02, "NetOps for the Global Information Grid (GIG)," Washington, DC: DoD Chief Information Officer, December 19, 2008.

23. Healey, p. 65.

24. "History of U.S. Strategic Command," from STRATCOM website, Offutt AFB, NE: U.S. Strategic Command, available from *www.stratcom.mil/history/*, accessed August 3, 2014, p. 3.

25. Healey, p. 66.

26. *Joint Concept of Operations for Global Information Grid NetOps*, pp. 17-18. Details of the JFCC-NW tasks included:

Network warfare is defined as the employment of Computer Network Operations (CNO) with the intent of denying adversaries the effective use of their computers, information systems, and networks, while ensuring the effective use of our own computers, information systems, and networks. This includes development of information/intelligence support and information assurance requirements for supporting network warfare, the integration of Computer Network Attack (CNA) and Computer Network Exploitation (CNE) capabilities and direct coordination with JTF-GNO.

27. Keith Alexander, "Warfighting in Cyberspace," *Joint Force Quarterly*, Issue 46, 3rd Quarter 2007, p. 61. Lieutenant General Alexander also noted that much work remained in the operationalization of cyberspace:

> While the concepts of NW and NetOps are a good start, they represent only a small subset of the elements of military power available within or enabled by cyberspace. In order to fully engage in the development of joint doctrine within the cyberspace domain, it is also necessary to develop a definition of exactly what warfare within cyberspace — or cyberspace warfare — is.

28. "Joint Task Force-Global Network Operations," *Index to Joint Enablers*, Handbook No. 10-60, Fort Leavenworth, KS: U.S. Army Combined Arms Center, August 2010, pp. 29-30.

29. Keith Alexander, Joint Functional Component Command for Network Warfare, Statement for the Record before the House Armed Services Committee on Terrorism, Unconventional Threats, and Capabilities Subcommittee, Washington, DC: U.S. House of Representatives, May 5, 2009, p. 1, available at *www.nsa.gov/public_info/speeches_testimonies/5may09_dir.shtml*, accessed on August 4, 2014.

30. William F. Lynn III, "Defending a New Domain: The Pentagon's Cyberstrategy," *Foreign Affairs*, Vol. 89, No. 5, September-October 2010, pp. 97-108.

31. Keith Alexander, "U.S. Cyber Command: Organizing for Cyberspace Operations," Hearing of Committee on Armed Services, House of Representatives, 111th Cong., 2d Sess., Washington, DC: U.S. Government Printing Office, September 23, 2010, p. 10. General Alexander explained the incident that drove the creation of Operation BUCKSHOT YANKEE:

> As I mentioned earlier, first, it became clear that we needed to bring together the offense and defense capabilities. And so Global Network Ops was put — Joint Task Force-Global Network Ops was put under my operational control in — within a month of that happening. And I think that started to change the way we look at this. And then the Secretary

of Defense set in motion the next step, which was to set up Cyber Command as a sub-unified command. And I think both of those are the right things to do. What it does is it gets greater synergy between those who are defending the networks and what they see and those that are operating in the networks abroad and what they see and bringing that together for the benefit of our defense. I think that is exactly what the Nation would expect of us.

The way that happens is, if you use a thumb drive or other removable media on an unclassified system, the malware would get on that removable media, ride that removable media over to the other system. And so think of it as a man in a loop wire, and so a person could be taking information they needed from an unclassified system, putting it onto a classified system, and so that software would ride that removable media and go back and forth. It [the malware] was detected by some of our network folks within the advanced network ops, our information assurance division at NSA.

32. Robert M. Gates, "Establishment of a Subordinate Unified U.S. Cyber Command under U.S. Strategic Command for Military Cyberspace Operations," memorandum for Secretaries of the Military Departments, Washington, DC: Office of the Secretary of Defense, June 23, 2009.

33. Alexander, "U.S. Cyber Command: Organizing for Cyberspace Operations," p. 6.

34. Gates, "Establishment of a Subordinate Unified U.S. Cyber Command under U.S. Strategic Command for Military Cyberspace Operations."

35. "Cyber Command Achieve Full Operational Capability," Department of Defense News Release No. 1012-10, Washington, DC: Department of Defense, November 3, 2010.

36. U.S. Cyber Command Factsheet, U.S. Strategic Command public website, Offutt AFB, NE: U.S. Strategic Command, August 2013, available from *www.stratcom.mil/factsheets/2/Cyber_Command/*, accessed August 3, 2013. The full mission statement is:

USCYBERCOM plans, coordinates, integrates, synchronizes and conducts activities to: direct the operations and defense of specified Department of Defense information networks and; prepare to, and when directed, conduct full spectrum military cyberspace operations in order to enable actions in all domains, ensure US/Allied freedom of action in cyberspace and deny the same to our adversaries.

Perhaps as an inside joke for the command's cryptologists, if you type in the USCYBERCOM mission statement verbatim into an md5 hash generator, the result will be "9ec4c12949a4f31474f-299058ce2b22a" which are the symbols that are written within the inner ring of the command's official seal.

37. "Defense Department Cyber Efforts: DoD Faces Challenge in its Cyber Activities," Report GAO-11-75, Washington, DC: U.S. Government Accountability Office, July 2011, p. 18.

38. Alexander, "U.S. Cyber Command: Organizing for Cyberspace Operations," pp. 38-41.

39. *Ibid.*, p. 8. In General Alexander's words:

What we have come up with is we need to set up a joint task force or, in this case, perhaps a joint cyber ops task force, and that cyber ops task force would work with Cyber Command, but go forward to work with the combatant command to present forces from all the services to meet in operational mission. And then let us train as a first step how each of those forces would do that, what we would do for PACOM [Pacific Command], CENTCOM [Central Command], EUCOM [European Command], SOUTH-COM [Southern Command], and NORTHCOM [Northern Command], if required.

40. *Ibid.*, p. 48. General Alexander went on to note:

The CSE supports the Combatant Commanders at their headquarters through liaison, planning, and operations support primarily at the Directorate of Operations, or J3 level. However, the CSE is empowered to develop relationships and capabilities across the Combatant Command. The CSEs have played innovative and complementary roles within the

COCOM Directorates of Intelligence (J2) and Directorates of Plans and Policy (J5). To enable their effectiveness, the CSE has full reach-back support to USCYBERCOM headquarters and the NSA Enterprise.

The size, composition, and role of an ExCSE team is scalable depending on mission requirements. For example, in Iraq and Afghanistan the ExCSEs provide cyber expertise directly to the deployed headquarters' planning effort while coordinating the delivery of cyber effects through USCYBERCOM headquarters and interagency partners. In future conflicts involving full-scale operations against sophisticated cyber adversaries, the ExCSEs will scale to meet mission requirements. The ExCSE teams will continue to coordinate for global effects through USCYBERCOM but will also play a key role in coordinating planning, direction, and execution of cyber operations through an in-theater Joint Cyber Operations Task Force (JCOTF).

41. *Ibid.*, pp. 10-11. General Alexander's argument for US-CYBERCOM's role in the unity of effort for cyberspace operations:

But the reality is, in cyberspace, that is — that is where NSA operates and has tremendous technical expertise. It has our Nation's expertise for crypto-mathematicians, for access, for linguists, for everything that you would need to operate in cyberspace.

And what the Secretary said is, we can't afford to replicate the hundreds of billions of dollars that we put into NSA to do another for Cyber Command and then another perhaps for DHS [Department of Homeland Security] and others. Let us leverage what we have and bring that together.

And so by bringing these two together, we have actually accomplished that goal. Now, they — they have and operate under separate staffs and under different auhorities, as you know. And so under the Cyber Command, the thing that has helped, I always had, since I have been the director of NSA, the additional duty as the Joint Functional Component Command-Net Warfare, so I had that job. What I didn't have was the staff, the — the horsepower and the staff that I have now, so actually that helps us.

42. *Ibid.*, p. 48.

43. Barack Obama and Leon Panetta, *Sustaining U.S. Global Leadership: Priorities for 21st Century Defense*, Washington, DC: Department of Defense, January 2012, p. 4. The priority statement reads:

> **Operate Effectively in Cyberspace and Space.** Modern armed forces cannot conduct high-tempo, effective operations without reliable information and communication networks and assured access to cyberspace and space. Today space systems and their supporting infrastructure face a range of threats that may degrade, disrupt, or destroy assets. *Accordingly, DoD will continue to work with domestic and international allies and partners and invest in advanced capabilities to defend its networks, operational capability, and resiliency in cyberspace and space.*

44. Rivers Johnson, "Command Overview Brief," Fort Meade, MD: U.S. Cyber Command, May 2012, slide 9, available from *www.defense.gov/DODCMSShare/briefingslide/363/Indonesian_Command_Briefing_Unclassified.pdf*, accessed on August 12, 2014. Details on the priorities were:
- Trained and Ready Cyber Forces
  - Increasing our capacity in numbers of personnel
  - Establishing joint training and certification standards
- Operational Concept
  - Describing how we will fight in cyberspace
  - Providing a model for unity of effort and unity of command
- Global Situational Awareness
  - Creating a cyber common operational picture
  - Enabling coordinated activities across the whole-of-government
- Defensible Architecture
  - Implementing a cloud-based, virtual single network
  - Ensuring secure, attribute-based access to data
- Policies and Procedures to Enable Action
  - Implementing Standing Rules of Engagement for cyber self-defense
  - Facilitating information sharing between government and industry

45. Jim Garamone, "Rogers Takes Over Top NSA, Cyber Command Posts," Washington, DC: DoD News, April 3, 2014. For reiteration of USCYBERCOM priorities, see Cheryl Pellerin, "Operationalizing Cyber is New Commander's Biggest Challenge," Washington, DC: DoD News, June 2, 2014; and Cheryl Pellerin, "Rogers: Cybercom Defending Networks, Nation," Washington, DC: DoD News, August 18, 2014.

46. Pellerin, "Operationalizing Cyber is New Commander's Biggest Challenge," p. 1. For details on the JIE, see "Joint Information Environment White Paper," Washington, DC: Joint Chiefs of Staff, January 22, 2013. The plan to have the first JIE structure in Europe is included on page 7:

> Over the next year [2013], we will begin to physically implement a JIE capable of supporting the needs of Joint Force 2020. Beginning in European and Africa Commands, followed by an incremental global rollout to the rest of the Joint Force, Joint Warfighters will have access to a common, protected information infrastructure with which to plan and fight together with our mission partners.

47. Department of Defense, *Quadrennial Defense Review 2014*, Washington DC: U.S. Government Printing Office, March 4, 2014, pp. 14-15. The 2014 QDR called for a total of 133 cyber teams to be available by fiscal year 2019:
    13 National Mission Teams with 8 National Support Teams
    27 Combat Mission Teams with 17 Combat Support Teams
    18 National Cyber Protection Teams (CPTs)
    24 Service CPTs
    26 Combatant Command and DoD Information Network CPTs

48. Pellerin, "Rogers: Cybercom Defending Networks, Nation," p. 1.

49. "Cyberspace Operations," USSTRATCOM Freedom of Information Act Reading Room document FOIA 14-003, Offutt AFB, NE: U.S. Strategic Command, October 8, 2013 (released February 26, 2014), slide 18, available from *www.stratcom.mil/files/foia_*

requests/FOIA%2014-003%20-%20Released.pdf, accessed August 21, 2014.

50. "Cyber Guard Exercise Tests People, Partnership," U.S. Cyber Command News Release, Fort Meade, MD: U.S. Cyber Command, July 17, 2014, p. 1. Exercise participants included:

> Elements of the National Guard, reserves, National Security Agency and U.S. Cyber Command exercised their support to Department of Homeland Security and FBI responses to foreign-based attacks on simulated critical infrastructure networks, promoting collaboration and critical information sharing in support of a "whole-of-nation" effort.

> The exercise also included several Cyber Protection Teams, part of Cybercom's Cyber Mission Force being built over the next few years. The teams defend DoD information networks and help support DoD's requirement to provide foreign intelligence and assessment and active-duty capabilities to defend the nation.

51. George J. Franz, III, "Effective Synchronization and Integration of Effect through Cyberspace for the Joint Warfighter," presentation at AFCEA TechNetLand Forces-East Conference in Baltimore, MD: Armed Forces Communications and Electronics Association, August 14, 2012, slide 12, available from *www.afcea. org/events/tnlf/east12/documents/4V3EffSynchIntEffthruCybrspcfor] tWarfighter_forpublicrelease.pdf*, accessed on August 21, 2014. The CERF process is described in the *JFIRE Multi-Service Procedures for the Joint Application of Firepower* (November 30, 2012), an unclassified but restricted distribution document available for authorized download from *www.alsa.mil/library/mttps/jfire.html*.

52. *Ibid.*, slide 18.

53. *Ibid.*, slide 9.

54. *Ibid.*, slide 10.

55. Joint Chiefs of Staff, Joint Publication 1-02, *Department of Defense Dictionary of Military and Associated Terms*, Washington, DC: Joint Chiefs of Staff, November 8, 2010 (as amended through

July 16, 2014), available from *www.dtic.mil/doctrine/new_pubs/ jp1_02.pdf,* accessed on August 21, 2012. JP 1-02 offers the following definitions:

> **defensive cyberspace operations** — Passive and active cyberspace operations intended to (a) preserve the ability to utilize friendly cyberspace capabilities and protect data, and (b) networks, net-centric capabilities, and other designated systems. Also called **DCO**. (JP 3-12)

> **defensive cyberspace operation response action** — Deliberate, authorized defensive measures or activities taken outside of the defended network to protect and defend Department of Defense cyberspace capabilities or other designated systems. Also called **DCO-RA**. (JP 3-12)

> **Department of Defense information networks** — The globally interconnected, end-to-end set of information capabilities, and associated processes for collecting, processing, storing, disseminating, and managing information on-demand to warfighters, policy makers, and support personnel, including owned and leased communications and computing systems and services, software (including applications), data, and security. (JP 3-12)

> **offensive cyberspace operations** — Cyberspace operations intended to project power by the application of force in or through cyberspace. Also called **OCO**. (JP 3-12)

56. Brett T. Williams, "The Joint Force Commander's Guide to Cyberspace Operations," *Joint Force Quarterly,* Issue 73, 2nd Quarter 2014, pp. 12-19.

57. Joint Doctrine Analysis Division, *Compendium of Key Joint Doctrine Publications,* Washington, DC: Deputy Directorate, Joint Staff, J-7, January 3, 2014, p. iii. The full synopsis states:

> JP 3-12, *Joint Cyberspace Operations,* is a new JP, signed 5 February 2013. It is classified SECRET and only resides on the SIPRNET JDEIS. It was initiated, based on the National Military Strategy for Cyberspace Operations Implementation Plan, which directed USSTRATCOM to assess joint doctrine in support of operations in cyberspace and the five National

Military Strategy Cyberspace Operations Ends. Initially a joint test publication (JTP), there was unanimous support by the joint doctrine development community to end development of the JTP and instead develop a JP.

58. "Network Enterprise Command Evolved from Strategic Communications Command," *Army Communicator* Vol. 35, No. 2, Summer 2010; PB 11-10-2, Fort Gordon, GA.: U.S. Army Signal Center, pp. 76-79.

59. Headquarters, Department of the Army, ESTABLISH-MENT OF U.S. ARMY NETWORK ENTERPRISE TECHNOL-OGY COMMAND/9th ARMY SIGNAL COMMAND; TRANS-FER AND REDESIGNATION OF THE HEADQUARTERS AND HEADQUARTERS COMPANY, 9th ARMY SIGNAL COM-MAND; DISCONTINUANCE OF THE COMMUNICATIONS ELECTRONIC SERVICES OFFICE AND THE INFORMATION MANAGEMENT SUPPORT AGENCY, General Order No. 5, Washington, DC: Headquarters, Department of the Army, August 13, 2002, p. 2.

60. *Joint Concept of Operations for Global Information Grid NetOps*, p. 25.

61. *Ibid.*, p. 30. Additional details of the Army NetOps structure circa 2005:

> The ANOSC is integrated with the 1st Information Operations Command (1st IO CMD – LAND) Army Computer Emergency Response Team (ACERT) to create a consolidated NetOps Center called ANOSC/ACERT Tactical Operations Center (A2TOC), and each TNOSC is integrated with a Regional Computer Emergency Response Team (RCERT). This alignment of organization has provided a critical synergism of effectiveness and efficiency to receive, distribute, and analyze information in order to integrate, synchronize, and coordinate CNO.

62. The term Army Network Operations and Security Center (ANOSC) was later changed to Army Global Network Operations and Security Center (AGNOSC), but it is not clear when this occurred since historical references uses them both in approximately the same time period from 2004-2006.

63. *Joint Concept of Operations for Global Information Grid NetOps*, p. 30.

64. *Ibid.*, p. 31.

65. Headquarters, Department of the Army, REINFORCING THE ESTABLISHMENT OF THE U.S. ARMY NETWORK EN-TERPRISE TECHNOLOGY COMMAND/9TH ARMY SIGNAL COMMAND AS A DIRECT REPORTING UNIT AND REDES-IGNATING THE COMMAND AS THE U.S. ARMY NETWORK ENTERPRISE TECHNOLOGY COMMAND/9TH SIGNAL COMMAND (ARMY), General Order No. 31, Washington, DC: Headquarters, Department of the Army, October 16, 2006, p. 1.

66. Floyd Light, "USASMDC/ARSTRAT's Joint Friendly Force Tracking Mission—Network Operations Success!" *Army Space Journal*, Vol. 8, No. 3, Fall 2009, p. 27.

67. *Ibid.*, p. 28.

68. *Ibid.*, pp. 26-31.

69. Headquarters, Department of the Army, "Cyber Opera-tions," 2009 U.S. Army Posture Statement Online Information Pa-per, Washington, DC: Department of the Army, available from *www.army.mil/aps/09/information_papers/cyber_operations.html*, ac-cessed on August 18, 2014. The information paper provided fur-ther details on Army NetWar units:

Army Network Operations (NetOps) forces assigned to the Army Network Enterprise Technology Command/9th Signal Command are stationed at forward locations with-in theater signal commands throughout each combatant commander's geographical area of responsibility (AOR). Network warfare (NetWar) forces assigned to the Army In-telligence and Security Command (INSCOM) are forward-based with theater military intelligence brigades in each of the combatant commander's AORs and integrated with NSA's worldwide operations. Army Information Opera-tions (IO) forces assigned to the 1st Information Operations Command (1st IO CMD) are deployed worldwide support-ing Joint and Army commanders with the planning, coordinating, integrating, and synchronizing of CNO capabilities into operational plans and orders.

In July 2008, the Army activated its first provisional NetWar Battalion under INSCOM. The mission of this battalion is to support both the Army and the DoD with a variety of tasks, ranging from tactical support to BCTs through strategic support for other Services, Joint commanders, and interagency partners.

The 1st IO CMD is a key component in integrating and synchronizing IO efforts, NetOps and NetWar capabilities with operational units through the global deployment of its support teams. In addition to supporting the A-GNOSC and TNOSCs with the Army Computer Emergency Response Team (CERT) and the Theater Regional CERTs respectively, 1st IO CMD provides critical cyberspace all-source intelligence support, testing of the network defenses, network forensic analysis, unit network vulnerability assessments, and CNO planning capabilities. The 1st IO CMD also provides Army cyber training support through its Basic CNO Planners Course, a newly approved Army skill identifier producing course.

70. "Establishment of U.S. Army Cyber Command," Army Cyber Command website available from *www.arcyber.army.mil/ history_arcyber.html*, accessed on August 18, 2014.

71. Headquarters, Department of the Army, "Cyber Operations," 2010 U.S. Army Posture Statement Online Information Paper, Washington, DC: Department of the Army, available from *https://secureweb2.hqda.pentagon.mil/VDAS_ArmyPostureStatement/2010/information_papers/Cyber_Operations.asp*, accessed on August 17, 2014. Further details on theater support included:

The Army provides the Combatant Commanders cyber support down to the tactical-level as part of DOD cyber operations. The TNOSC and the Regional Computer Emergency Response Teams (RECERT) provide support to the COCOMs through six locations that have complete and overlapping coverage of the COCOM Area of Responsibility (AOR).

72. Rhett Hernandez, U.S. Army Forces Cyber Command, Statement before the House Armed Services Committee Subcommittee on Terrorism, Unconventional Threats and Capabilities, 111th Cong., 2nd Sess., Washington, DC: U.S. House of Representatives, September 23, 2010, p. 4.

73. *Ibid.*, p. 6.

74. *Ibid.*, p. 7. Details of the Army liaison with USCYBERCOM include:

> To ensure the ACOIC is fully nested with and able to seamlessly support USCYBERCOM, the ACOIC is physically locating and embedding approximately 25 personnel in the USCYBERCOM joint staff. These embedded personnel will ensure close collaboration with USCYBERCOM and enable the ACOIC to leverage USCYBERCOM's unique resources and capabilities.

75. *Ibid.*, p. 8.

76. Department of the Army, TRADOC Pamphlet 525-7-8, *Cyberspace Operations Concept Capability Plan 2016-2028*, Fort Monroe, VA: Headquarters, U.S. Army Training and Doctrine Command, February 22, 2010, p. ii.

77. *Ibid.*, p. i.

78. *Ibid.*

79. *Ibid.*, p. 17. Pages 18-22 of the pamphlet contain the following definitions for these four components:

> (p. 18) CyberSA is the immediate knowledge of friendly, adversary and other relevant information regarding activities in and through cyberspace and the EMS. It is gained from a combination of intelligence and operational activity in cyberspace, the EMS, and in the other domains, both unilaterally and through collaboration with unified action and public-private partners. Discrimination between natural and manmade threats is a critical piece of this analysis. CyberSA enables informed decisionmaking at all levels via flexibly tailored products and processes that can range from broadly

disseminated awareness bulletins targeted to general users to the other extreme of specific and narrowly focused issues distributed as extremely sensitive and classified in nature. CyberSA enables informed decisionmaking at all levels. It is relevant at the strategic, operational, and tactical echelons for overall SA; and it is useful to Soldiers who interact most with the populace, which uses and increasingly relies on cyberspace.

(p.19) CyNetOps is the component of CyberOps that establishes, operates, manages, protects, defends, and commands and controls the LandWarNet14, critical infrastructure and key resources (CIKR), and other specified cyberspace. CyNetOps consists of three core elements: Cyber enterprise management (CyEM), cyber content management (CyCM), and cyber defense (CyD), including information assurance, computer network defense (to include response actions), and critical infrastructure protection. CyNetOps uses CyEM, CyCM, and CyD in a mutually supporting and supported relationship with CyberWar and CyberSpt.

(p. 21) CyberWar is the component of CyberOps that extends cyber power beyond the defensive boundaries of the GIG to detect, deter, deny, and defeat adversaries. CyberWar capabilities target computer and telecommunication networks and embedded processors and controllers in equipment, systems and infrastructure. CyberWar uses cyber exploitation (CyE), cyber attack (CyA), and dynamic cyber defense (DCyD) in a mutually supporting and supported relationship with CyNetOps and CyberSpt.

(p. 22) CyberSpt is a diverse collection of supporting activities which are generated and employed to specifically enable both CyNetOps and CyberWar. These activities are called-out in this unifying category due to their unique and expensive nature as high-skilled, low-density, time-sensitive/intensive activities requiring specialized training, processes, and policy. Additionally, several of these activities also require specialized coordination, synchronization, and integration to address legal and operational considerations. It is because of these considerations and their overall importance that these activities are addressed as a CyberOps core component.

80. *Ibid.*, pp. 46-57.

81. Headquarters, Department of the Army, ESTABLISH-
MENT OF THE UNITED STATES ARMY CYBER COMMAND,
General Order No. 2010-26, Washington, DC: Headquarters, De-
partment of the Army, October 1, 2010.

82. Rhett Hernandez, Commanding General U.S. Army Cyber
Command/2nd Army, "Concerning Digital Warriors: Improving
Military Capabilities in the Cyber Domain," Statement before the
House Armed Services Committee Subcommittee on Emerging
Threats and Capabilities, 112th Cong., 2nd Sess., Washington,
DC: U.S. House of Representatives, July 25, 2012, p. 2.

83. General Order No. 2010-26, p. 1. The mission statement for
ARCYBER was:

> Army Cyber Command is the lead for Army missions, ac-
> tions and functions related to cyberspace, and responsible
> for planning, coordinating, integrating, synchronizing, di-
> recting and conducting Army network operations and the
> defense of all Army networks. When directed, Army Cyber
> Command conducts a full range of cyberspace operations
> to ensure freedom of action in cyberspace, and to deny the
> same to our adversaries. Army Cyber Command serves as
> the single Army point of contact for reporting and assessing
> Army cyberspace incidents, events, and operations and for
> synchronizing and integrating responses thereto.

84. John M. McHugh, Secretary of the Army Memorandum,
Subject: Army Directive 2011-03 (Change of Operational Control
for 1st Information Operations Command (Land) and Direction
for U.S. Army Cyber Command to Conduct the Information Operations Missions for the Army), Washington, DC: Headquarters,
Department of the Army, February 2, 2011, p. 1. The directive describes the mission as:

> The 1st Information Operations Command (Land) is the
> Army's only active duty command providing full-spectrum
> IO support. Its mission is to provide IO support to the Army
> and other military forces through deployable IO support
> teams; IO reachback planning and analysis, and the syn-
> chronization and conduct of computer network operations

(CNO); and in conjunction with other CNO and network operations stakeholders, to operationally integrate IO, reinforce forward IO capabilities, and to defend cyberspace in order to enable IO throughout the information environment.

85. Headquarters, Department of the Army, "Cyber Operations," 2012 U.S. Army Posture Statement Online Information Paper, Washington, DC: Department of the Army, available from *https://secureweb2.hqda.pentagon.mil/VDAS_ArmyPostureStatement/2012/InformationPapers/ViewPaper.aspx?id=192*, accessed on August 18, 2014.

86. Hernandez, "Concerning Digital Warriors: Improving Military Capabilities in the Cyber Domain," p. 6. For additional background details regarding the establishment of the cyber brigade, see "Establishment of U.S. Army Cyber Command," Army Cyber Command website:

In addition to naming the SMDC/ARSTRAT as the interim ARFORCYBER, the execute order [Department of the Army G-3/5/7 Execute Order (No. 155-10), released on May 11, 2010] directed the G-2 to develop a plan for an Army Cyber Brigade, subordinate to INSCOM, to conduct Signals Intelligence, Computer Network Operations and Dynamic Cyber Defense in support of ARFORCYBER-directed operations. The 744th Military Intelligence Battalion (formerly Army Network Warfare Battalion) of the 704th Military Intelligence Brigade currently performed the mission. The Army subsequently approved the establishment of an Army cyber brigade in December 2010 and designated the 780th Military Intelligence Brigade, with an effective date of Oct. 1, 2011, to fulfill the 744th MI Battalion's mission.

87. "Establishment of U.S. Army Cyber Command," p. 7.

88. Rhett Hernandez, "Army Cyber Command/Second Army," PowerPoint presentation at LandWarNet 2011 conference, Tampa, FL: Armed Forces Communications and Electronics Association, August 25, 2011, slide 3, available from *www.afcea.org/events/pastevents/documents/H325Aug.pdf*, accessed on August 18, 2014.

89. *Ibid.*, slide 4. The nine accomplishments were (note that bolded words in text are from the presentation slide):

- **Integration with US Cyber Cmd and Service Components**
- **Operational Focus** with **Unprecedented Unity of Effort** in Operating and Defending All Army Networks
- Significant Contribution to **Operational Cyber Planning**
- Increased **Full Spectrum Capacity and Capability**
- Continued Focus on **Enterprise Capabilities**
- **Integrated Cyberspace Opns** [operations] in Major CO-COM [combatant command] exercises
- Established a **Army Cyberspace Proponent Office**
- Aggressively Developing **Cyberspace Requirements**
- Producing **"Army Cyber 2020" Strategic Plan**

90. Headquarters, Department of the Army, "Cyberspace: Army Cyber Command and Cyberspace Operations," 2012 U.S. Army Posture Statement Online Information Paper, Addendum K, Washington, DC: Department of the Army, available from *https://secureweb2.hqda.pentagon.mil/VDAS_ArmyPostureStatement/2012/InformationPapers/ViewPaper.aspx?id=192*, accessed on August 19, 2014. The full text of the three cyberspace elements for 2020:

> A cyberspace enterprise that provokes the shift from services provided by Network Operations to the network as a war fighting platform, with a cyberspace infrastructure that enables maneuver through the hostile, competitive environment of cyberspace.
>
> A "combined arms" cyberspace force of elements that conduct the full range of cyberspace operational functions (build, operate, defend, exploit, and attack) to support friendly effects and counter adversarial advances.
>
> The integration, planning, and synchronization of cyberspace effects with all warfighting functions and across all domains, delivering effects to support commander's objectives.

91. *Ibid.*, p. 2.

92. *Ibid.*

93. *Ibid.*

94. Hernandez, "Concerning Digital Warriors: Improving Military Capabilities in the Cyber Domain," p. 6.

95. "Moving to the Future: Exclusive Interview with Lt. Gen. Edward C. Cardon," *Army Technology Magazine*, Vol. 1, No. 2, October 2013, p. 4.

96. Headquarters, Department of the Army, AFFIRMATION OF SECRETARY OF THE ARMY COMMITMENT TO UNITY OF EFFORT; DESIGNATION OF U.S. ARMY CYBER COMMAND AS AN ARMY FORCE COMPONENT HEADQUARTERS; REACTIVATION OF SECOND ARMY AND DESIGNATION AS A DIRECT REPORTING UNIT; DISESTABLISHMENT OF THE U.S. ARMY NETWORK ENTERPRISE TECHNOLOGY COMMAND/9TH SIGNAL COMMAND (ARMY) AS A DIRECT REPORTING UNIT AND REASSIGNMENT TO SECOND ARMY; DESIGNATION OF GENERAL COURT-MARTIAL CONVENING AUTHORITIES, General Order No. 2014-02, Washington, DC: Headquarters, Department of the Army, March 6, 2014.

97. Richard A. Davis, "What's Cyber Hot?" PowerPoint presentation at ISACA Conference DoD Training Day, Arlington, VA: ISACA, February 11, 2014, slide 4, available from *www.isaca-washdc.org/presentations/2013/201402-session5.pdf*, accessed on August 19, 2014.

98. Edward C. Cardon, Keynote Address at "Securing America's Future in the New 'Interwar Years'," 5th Annual Military and Federal Fellow Research Symposium, Washington, DC: The Brookings Institution, March 12, 2014, p. 59.

99. *Ibid.*, p. 66.

100. Hernandez, "Concerning Digital Warriors: Improving Military Capabilities in the Cyber Domain," p. 6. Additional details on the support provided by Army Reserve forces:

This support includes conducting IO and cyberspace operations planning, preparation, execution and assessment of the information environment; identifying IO and cyberspace vulnerabilities; leveraging IO and cyberspace intelligence analysis; and conducting training in IO and cyberspace operations to improve a unit's ability to successfully operate throughout the information environment.

101. 2015 National Guard Bureau Posture Statement, *Trusted at Home, Proven Abroad*, Washington, DC: National Guard Bureau, p. 38. Advantages of using National Guard cyberspace forces cited in the statement include:

- As the cyber mission continues to be defined, grown and matured, the National Guard remains uniquely positioned to provide a cost-effective and capable force to support the DoD, homeland defense and civil support activities.
- Unique legal authorities when on state active duty, allow our Soldiers and Airmen to work with law enforcement during a cyber emergency. Additionally, our knowledge of local critical infrastructure vulnerable to cyber attack, combined with longstanding relationships with those owners and operators, enables us to respond quickly.
- Working for renowned IT companies, our Guard members possess unique civilian skills. The Guard's part-time structure also helps recruit and retain patriotic and skilled cyber warriors who want to serve their country. Currently, our Soldiers and Airmen support a wide range of federal and state cyber missions.

Also, the report provides five specific examples of National Guard units that currently support cyberspace operations:

- Army Guard cyber specialists operate the Data Processing Unit / Information Operations Support Center.
- Army Guard Computer Network Defense Teams control and operate the defensive cyber system that is embedded in every state, territory and District of Columbia National Guard headquarters.
- A full-time Army Guard Cyber Protection Team (CPT) is currently being established to defend and secure DoD infrastructure and protect DoD networks.
- The Air Guard currently has 5 Network Warfare Squadrons, 3 Information Operation Squadrons, and 1 Information Aggressor Squadrons providing an array of critical cyber capabilities.

- Five Air Guard cyber units (3 operational, 2 in conversion) are providing intelligence that helps identify those trying to exploit computer networks.

102. Mike Milord, "Army Cyber Command, Army Guard sign memorandum to integrate cyber protection team," Fort Belvoir, VA: Army Cyber Command, June 5, 2014, available from *www. army.mil/article/127442/Army_Cyber_Command__Army_Guard_ sign_memorandum_to_integrate_cyber_protection_team/?from=RSS 1/*, accessed on August 19, 2014.

103. "Army Announces Decision of Army Cyber Forces," Department of Defense News Release No. NR-084-13, Washington, DC: Department of Defense, December 19, 2013.

104. *Ibid.*, p. 1.

105. Russell Fenton and David L. Smith, "Signal Center Changes to Cyber Center" and "New Mission Integrates Related Cyberspace Operations Training," *Army Communicator* Vol. 39, No. 1, Spring 2014, pp. 7-10.

106. Edie M. Fairbank, "Cyber Center of Excellence Generates Need for New Doctrine," *Army Communicator* Vol. 39, No. 1, Spring 2014, pp. 11-12. When completed, FM 3-12, *Cyberspace Operations*, is expected to address:

This manual links joint cyberspace operations doctrine and ADRP 3-0, Unified Land Operations, providing the methods by which Army forces support and perform offensive cyberspace operations, defensive cyberspace operations, and Department of Defense information network operations, providing opportunities for commanders to integrate specialized cyberspace capabilities in support of their concept of operations. This manual provides an overview of cyberspace and its relationship to the operational environment; examines the roles, responsibilities, and working relationships of joint and Army cyber organizations involved in cyberspace operations; and discusses how cyberspace operations are an integral part of unified land operations. Supporting cyber ATPs are in the planning phase of development.

107. Department of Army, *FM 3-38, Cyber Electromagnetic Activities*, Washington, DC: Headquarters, Department of Army, February 2014, p. v.

FM 3-38 is the first doctrinal field manual of its kind. The integration and synchronization of cyber electromagnetic activities (CEMA) is a new concept. The Army codified the concept of CEMA in *Army Doctrine Publication (ADP) 3-0, Unified Land Operations*, and ADP 6-0, *Mission Command*. The mission command warfighting function now includes four primary staff tasks: conduct the operations process (plan, prepare, execute, assess), conduct knowledge management and information management, conduct inform and influence activities (IIA), and conduct CEMA.

108. Johnson, slide 4.

109. *Cyber Electromagnetic Activities*, p. 3-2.

110. Davis, slide 7.

111. *Ibid.*, slide 6.

112. Siobhan Carlile, "Army Recruiting Highly Qualified Soldiers, DA Civilians to Serve on New Specialized Cyber-Protection Teams," *Army Communicator*, Vol. 38, No. 3, Fall 2013, pp. 8-9.

113. Jeffrey L. Caton, "Examining Active Cyber Defense in Deterrence and Conflict Escalation," presentation at Cyber Security Day, Session 9-Security trends and solutions, Prague, Czech Republic: Future Forces Exhibition and Conference International, October 17, 2014. The paper is a modified version of this section of this monograph.

114. *Sustaining U.S. Global Leadership*, p. 7.

115. "Advance Questions for Vice Admiral Michael S. Rogers, USN Nominee for Commander, United States Cyber Command," Washington, DC: U.S. Senate, March 11, 2014, p. 17. In his statement he also provided the following information regarding cyberspace and deterrence:

(Congressional question) Would you agree that promulgating such a doctrine requires at least some broad statements of capabilities and intentions regarding the use of offensive cyber capabilities, both to influence potential adversaries and to reassure allies?

(Admiral Rogers response) Classic deterrence theory is based on the concepts of threat and cost; either there is a fear of reprisal, or a belief that an attack is too hard or too expensive. Cyber warfare is still evolving and much work remains to establish agreed upon norms of behavior, thresholds for action, and other dynamics. A broad understanding of cyber capability, both defensive and offensive, along with an understanding of thresholds and intentions would seem to be logical elements of a deterrence strategy, both for our allies and our adversaries and as they are in other warfighting domains. I believe we'll see much discussion of the structure and implementation of our cyber deterrence strategy from DoD and Intelligence Community experts, along with Interagency engagement. (pp. 17-18.)

116. Hernandez, "Concerning Digital Warriors: Improving Military Capabilities in the Cyber Domain," p. 5.

The Army Cyber Defense in Depth strategy (Active Defense) facilitates a clear identification and prioritization of key cyber terrain, including physical and logical infrastructure and mission data. The strategy employs three overarching strategic objectives to protect key cyber terrain: Protect, including Defense of the Global Information Grid Operations (DGO) and Information Assurance (IA) measures; Defend, including passive Defensive Cyber Operations (DCO) organized around the deployment of perimeter and key terrain focused sensors, firewalls, and various host-based security systems and programs; Hunt, consisting primarily of active DCO utilizing advanced "active" sensors and rapid response actions. We continue to increase our capacity and capability to conduct each objective and our efforts will remain synchronized with the transition to the DOD Joint Information Environment (JIE).

117. James R. Gosler and Lewis Von Thaer, "Task Force Report: Resilient Military Systems and the Advanced Cyber Threat," Washington, DC: Department of Defense, Defense Science Board, January 2013, pp. 6-15.

118. Richard L. Kugler, "Deterrence in Cyber Attacks" F. Kramer, S. Starr, and L. Wentz, eds., *Cyberpower and National Security*, Washington, DC: Potomac Books, 2009, pp. 309-340.

119. *Joint Publication 3-0: Joint Operations*, Washington, DC: Joint Chiefs of Staff, August 11, 2011.

120. *Deterrence Operations Joint Operating Concept*, Version 2.0., Washington, DC: DoD, December 2006, pp. 3-7.

121. Herman. Kahn, *On Escalation: Metaphors and Scenarios*, New York: Praeger, 1965, p. 37.

122. *Ibid.*, pp. 6, 38.

123. Gosler and Von Thaer, pp. 40-45.

124. Kahn, pp. 45-46.

125. *Ibid.*, p. 39.

126. *Ibid.*, pp. 41-45.

127. *Ibid.*, pp. 46-51.

128. *Ibid.*, pp. 7-9, 15-23.

129. *Ibid.*, pp. 217, 220, 231.

130. Myriam Dunn Cavelty, "The Reality and Future of Cyberwar," Zurich, Switzerland: CSS Analysis in Security Policy, 2010.

131. Martin Libicki, *Crisis and Escalation in Cyberspace*, Santa Monica, CA: RAND Corporation, 2012.

132. *Department of Defense Strategy for Operating in Cyberspace*, Washington DC: DoD, July 2011.

133. *Capstone Concept for Joint Operations: Joint Force 2020*, Washington, DC: Joint Chiefs of Staff, September 10, 2012.

134. Martin Dempsey, "Defending the Nation at Network Speed: A Discussion on Cybersecurity with General Martin E. Dempsey, U.S. Army," presentation transcript, Washington, DC: The Brookings Institution, June 27, 2013.

135. Gosler and Von Thaer, pp. 8-9, 40-45.

136. Kahn, pp. 5, 9.

137. *Ibid.*, pp. 9-13.

138. *Department of Defense Strategy for Operating in Cyberspace*, p. 7.

139. Dempsey, pp. 42-43.

140. Michael Schmitt, "Attack as a Term of Art in International Law: The Cyber Operations Context," *4th International Conference on Cyber Conflict Proceedings*, Tallinn, Estonia: NATO Cooperative Cyber Defence Center of Excellence, 2012, pp. 283-293.

141. Dempsey, p. 42.

142. Forest Hare, "The Significance of Attribution to Cyberspace Coercion: A Political Perspective," *4th International Conference on Cyber Conflict Proceedings*, Tallinn, Estonia: NATO Cooperative Cyber Defence Center of Excellence, 2012, pp. 125-139.

143. Jeffrey L. Caton, "Exploring the Prudent Limits of Automated Cyber Attack," K. Podins, J. Srinissem, and M. Maybaum, eds., *5th International Conference on Cyber Conflict Proceedings*, Tallinn, Estonia: NATO Cooperative Cyber Defence Center of Excellence and IEEE, 2013, pp. 145-160.

144. Dempsey, p. 42.

145. Kahn, p. 8.

146. Selmer Bringsjord *et al.*, "Nuclear Deterrence and the Logic of Deliberate Mindreading," Cognitive Systems Research, June 23, 2013, available from *kryten.mm.rpi.edu/SB_NSG_SE_EM_ JL_nuclear_mindreading_062313.pdf*, accessed August 22, 2014.

147. Gosler and Von Thaer, pp. 8, 42-44.

148. Barack Obama, *International Strategy for Cyberspace: Prosperity, Security, and Openness in a Networked World*, Washington, DC: The White House, May 2011, pp. 10, 14.

149. Jason Healey and A. J. Wilson, "Cyber Conflict and the War Powers Resolution: Congressional Oversight of Hostilities in the Fifth Domain," Washington, DC: The Atlantic Council, 2013.

150. Hardin Tibbs, "The Global Cyber Game: Achieving Strategic Reliance in the Global Knowledge Society," Shrivenham, UK: The Defence Academy of the United Kingdom, pp. 50-52.

151. Colin Gray, *Making Strategic Sense of Cyber Power: Why the Sky is not Falling,* Carlisle, PA: Strategic Studies Institute, U.S. Army War College, April 2013, pp. ix-xi, 44-54.

152. Kugler, p. 324.

153. Eric Talbot Jensen, "Cyber Deterrence" *Emory International Law Review*, Vol. 26, Issue 2, 2012, pp. 773-824.

154. Vincent Manzo, "Deterrence and Escalation in Cross-Domain Operations: Where Do Space and Cyberspace Fit?" Washington, DC: National Defense University, December 2011, p. 7.

155. Gosler and Von Thaer, p. 8.

156. Jensen, pp. 773-824.

157. Manzo, p. 5.

158. Melissa Hathaway, "Cyber Readiness Index 1.0.," Great Falls, VA: Hathaway Global Strategies LLC, 2013, available from *belfercenter.hks.harvard.edu/files/uploads/Cyber-Readiness-Index-1-0-November-2013.pdf*, accessed August 22, 2014.

159. *Cyberspace Operations Concept Capability Plan 2016-2028*, p. i.

160. Ronald J. Deibert, *Bounding Cyber Power: Escalation and Restraint in Global Cyberspace,* Internet Governance Paper No. 6. Ontario, Canada: The Centre for International Governance Innovation, October 2013, p. 12.

161. *Cyberspace Operations Concept Capability Plan 2016-2028,* p. i.

162. Kugler, pp. 336-337.

163. Sean Lawson, "Putting the 'War' in Cyberwar: Metaphor, Analogy, and Cybersecurity Discourse in the United States," First Monday: Peer-Reviewed Journal on the Internet, Vol. 17, No. 7, July 2, 2012, available from *firstmonday.org/ojs/index.php/fm /article/view/3848/3270,* accessed August 22, 2014.

164. Deibert, p. 10.

APPENDIX

The following diagram is taken from Chapter IV of *JP 3-12(R), Cyberspace Operations*, that was declassified and posted for public access on October 21, 2014. It depicts typical military cyberspace command and control structures for steady-state and contingency operations. Note that the organization listed as "USS-RATCOM" in the upper left corner of the figure is a typographic error for "USSTRATCOM."

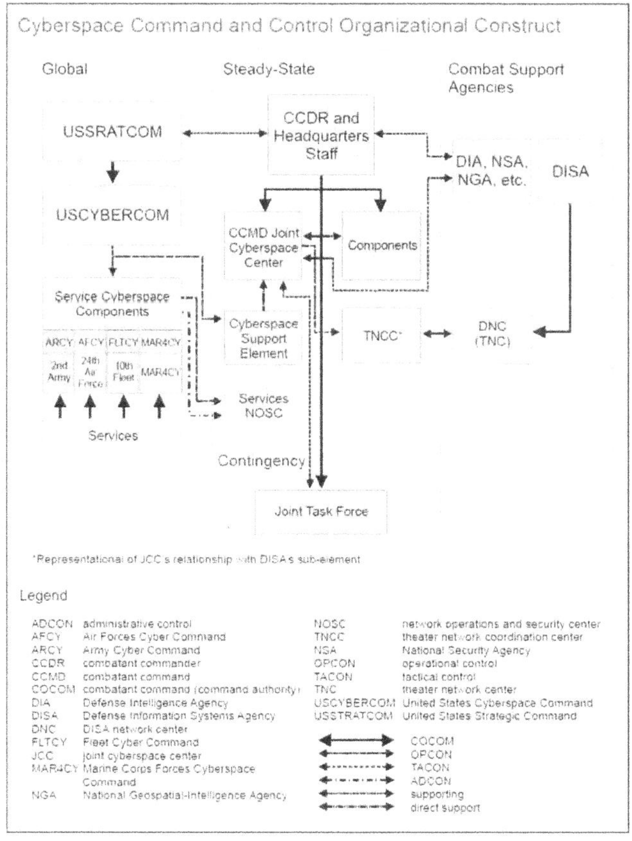

Figure A-1. Cyberspace Command and Control Organizational Construct.

U.S. ARMY WAR COLLEGE

Major General William E. Rapp
Commandant

*****

STRATEGIC STUDIES INSTITUTE
and
U.S. ARMY WAR COLLEGE PRESS

Director
Professor Douglas C. Lovelace, Jr.

Director of Research
Dr. Steven K. Metz

Author
Mr. Jeffrey L. Caton

Editor for Production
Dr. James G. Pierce

Publications Assistant
Ms. Rita A. Rummel

*****

Composition
Mrs. Jennifer E. Nevil

Made in the USA
Monee, IL
07 July 2026

56551502R00059